JOHN STROHMEYER

EXTREME CONDITIONS

Big Oil and the Transformation of Alaska

SIMON & SCHUSTER
New York London Toronto Sydney Tokyo Singapore

SIMON & SCHUSTER
Simon & Schuster Building
Rockefeller Center
1230 Avenue of the Americas
New York, New York 10020

Designed by Deirdre C. Amthor

Manufactured in the United States of America

1 3 5 7 9 10 8 6 4 2

Library of Congress Cataloging-in-Publication Data

Strohmeyer, John, date.
Extreme conditions : big oil and the transformation of Alaska /
John Strohmeyer.
p. cm.
Includes index.
1.Petroleum industry and trade—Alaska. 2.Petroleum industry
and trade—Environmental aspects—Alaska.
I. Title.
HD9567.A4S77 1993
338.2'7282'09798—dc20
93-16573
CIP
ISBN 0-671-76697-X

Contents

	Preface	9
1	An Awakening	13
2	The New Gold Rush	23
3	The First Big Strike	35
4	Race to the Arctic	48
5	The Settlement	61
6	Birth of the Pipeline	76
7	Pipe Dreams and Schemes	86
8	Days of Milk and Barley	99
9	Season of Gold	110
10	Corruption, Alaska Size	123
11	Anchorage Meets Dallas	151
12	The Iron Fist of Jesse Carr	162
13	Eskimo Capitalists	175
14	Here Is Free	190
15	Power Politics	205
16	Hard Aground	217
17	Mopping Up	227
18	Day of Reckoning	241
19	The New Battle	255
	Acknowledgments	271
	Index	275

ARCTIC OCEAN
Barrow
SIBERIA
Chukchi Sea
Umiat
BROOKS RANGE
Kotzebue
Nome
Yukon R.
Bering Sea
ALASKA
Bethel
KODIA
ISLAND
Bristol Bay
Kms.
0
200
0
200
Miles

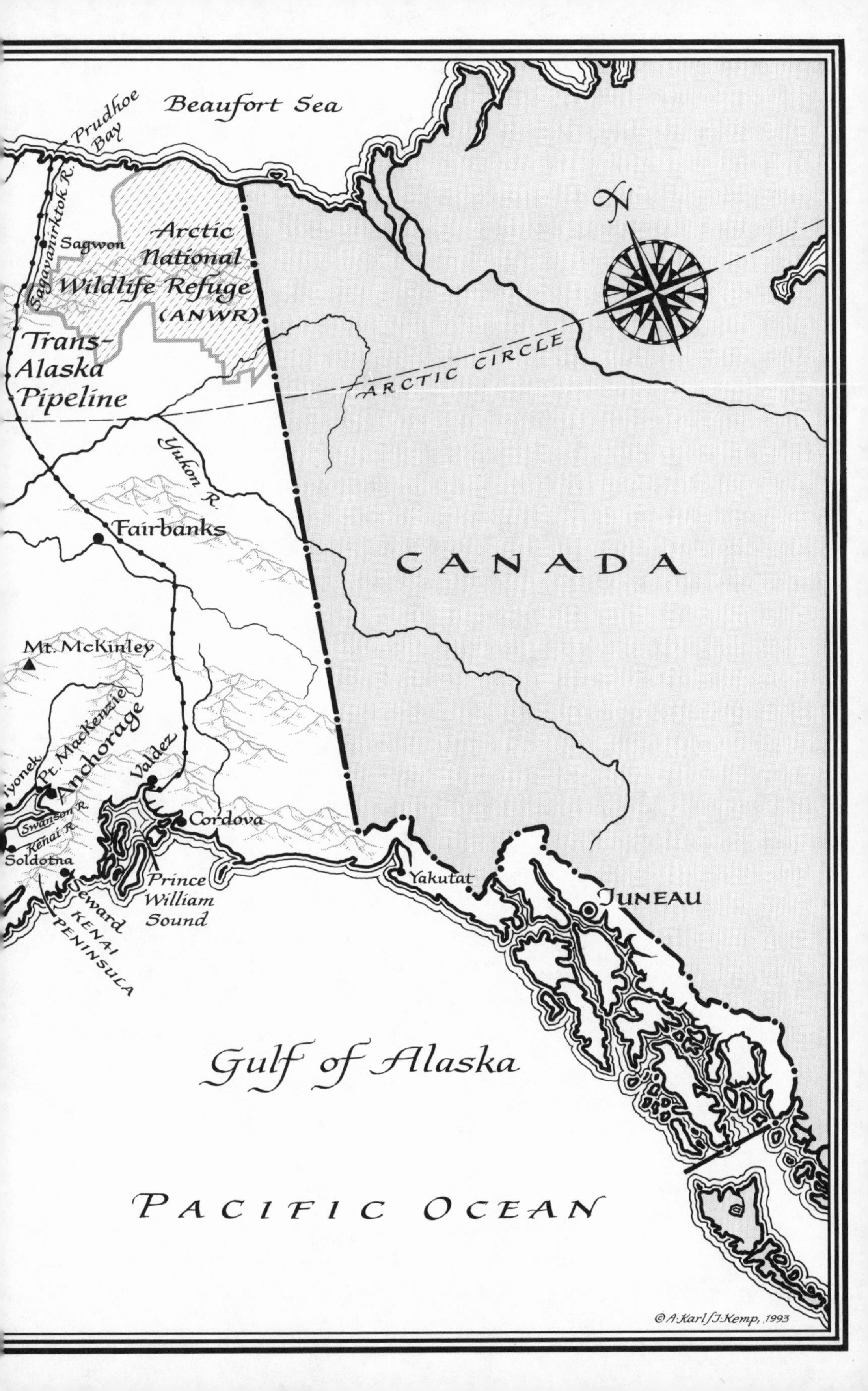
Beaufort Sea
Prudhoe Bay
Sagavanirktok R.
Sagwon
Arctic National Wildlife Refuge (ANWR)
Trans-Alaska Pipeline
ARCTIC CIRCLE
N
Yukon R.
Fairbanks
CANADA
Mt. McKinley
Pt. Mackenzie
Anchorage
Valdez
Swanson R.
Kenai R.
Soldotna
Cordova
Seward
Prince William Sound
KENAI PENINSULA
Yakutat
JUNEAU
Gulf of Alaska
PACIFIC OCEAN
©A·Karl/J·Kemp, 1993

Preface

This is a book about a different Alaska. You will find no scare stories about people seized with cabin fever and struggling to hold on until breakup thaws their roof-high snowdrifts. Don't look for gun battles outside the Last Chance Saloon as miners try to settle claim-jumping feuds. Nor will you read about grizzlies attacking sourdoughs, hunters, fishermen, hikers, tourists, authors, or any of the others who venture into bear country for myriad reasons.

All these events have occurred in Alaska past, and many still do. But this is a book about the more significant perils confronting this great land today—specifically, the threats to the face and character of Alaska that have been growing since the largest oil field in North America was discovered here twenty-five years ago.

I came to Alaska in 1987 with no intention of writing a book. Having chosen to retire early from a career of daily newspaper deadlines, I set out to do what I'd always dreamed of doing—tour this majestic land and fish its famed salmon and rainbow trout waters. A successful bid for an endowed chair as professor of journalism at the University of Alaska Anchorage gave me my first base in the state. After two years of teaching, listening to Alaskans wherever I met them, and fascinated reading of the state's newspapers, I decided to give up teaching, cut down on my fishing, and try to tell a story of Alaska that I felt needed to be told.

I spent three more years doing research. I drove every road that led to any village. Bush pilots flew me to many of the remote settlements that roads did not connect. I sought the confidence of government officials, beat reporters, and Natives who had front-row seats during the tumultuous years of change in Alaska.

I expected to find many inspiring stories. Certainly the daring and expertise of the men who found oil after a half century of fruitless exploration in this forbidding state is one of them. Alaska Natives, perennially struggling to live off the land and the sea, stood to benefit from the discovery. And indeed they won lavish land claims settlements, exceeding all that the nation had ever paid to the displaced American Indians in the lower forty-eight.

But I soon found that outright grants of cash and land do not guarantee freedom from economic distress, and, in fact, seem to invite new problems. Native elders see a link to increasing social problems and a decline in family bonding as subsistence life-styles give way to dependence on a cash economy. "Consider the choices for our youth," a teacher in the Eskimo village of Bethel laments. "Remain in a village with a fading culture or strike out into a world where they are unprepared to survive."

The troubles of these Native people were but one tremor from the massive eruption of oil money in the state. The giant oil field at Prudhoe Bay generated more than $110 billion in oil money since pumping started in 1978. Counting revenues from lease sales, more than $35 billion has gone directly into the Alaska treasury—an awesome windfall for a state that before the discovery had fewer than 250,000 residents and an average annual budget of less than $1 billion.

This torrent of money ignited a reckless state spending binge, unprecedented in this nation's history. Many of the expenditures improved Alaska's primitive infrastructure, and the state had the foresight to create a permanent savings fund. But along with the benefits came abuses. Alaska became a feasting ground for ambitious and often naive empire builders. Bribe-paying developers, greedy labor bosses, and professional racketeers thrived while the boom lasted. Lawyers, bankers, and crafty carpetbaggers exploited the riches without shame. Corruption flourished, the rule of law suffered, the wilderness shrank. Intoxicated by oil revenues, the state lost sight of its role as a regulator of the industry that was enriching it. The oil companies took full advantage. They rearranged the legislature to their liking with hefty campaign contributions. Only after the largest oil spill in the nation's history occurred in Alaska did the extent of this unholy alliance begin to surface.

The exploitation of Alaska is one concern of this book. Another is the bitter emotional issues tearing apart the state. As oil revenues decline, an aggressive and sometimes lawless governor heats up the conflict between states' rights and federal policy over developing more of Alaska's pristine territories. As game and salmon decrease, urban, Native, and commercial coalitions battle over who has priority to hunt and fish. And there remains the plight of the Natives, struggling to succeed as capitalists while they try to hold on to an ancient culture.

I have attempted to name the exploiters and the exploited, the corrupters and the caretakers, the betrayers of public trust and the battlers for justice. My intention has been to provide an insight into the forces that have been changing Alaska, and an understanding of the stakes involved in the intensifying tug-of-war over the future direction of this nation's unique and truly Great Land.

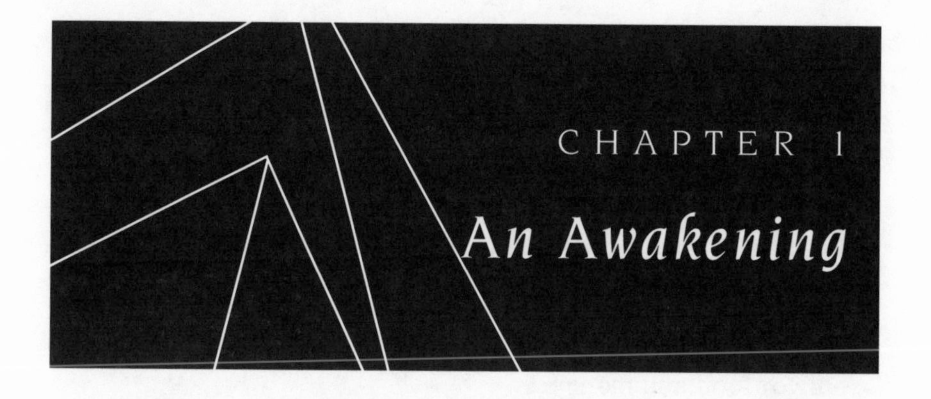

CHAPTER 1
An Awakening

ONE DAY in late April 1953, Willie Hensley, a twelve-year-old Inupiat Eskimo, awoke on his bed of willow twigs in his family's sod hut in the little settlement of Itakatuq, 28 miles above Alaska's Arctic Circle, and felt a change in the weather.

"Kiita! Kiita!" shouted Priscilla Hensley, the matriarchal leader, rousing the others. "Let's go!"

It was a day for the males to work on replenishing the family's fading cash supply. A warm storm out of the west, known as a chinook, had swept away the arctic spring's subfreezing temperatures and brought in a balmy thirty-five degrees. It was ideal weather for setting traps in the muskrat push-ups that protrude above the surface of the still-frozen sloughs and lakes.

Willie had been taught by his elders that as winter softens, muskrats make frequent trips to these little feeder houses on the top of the ice. There they sit on the edge of the hole to feed on vegetation that they bring up from below. As trappers did generations before him, Willie cut into the sides of the little clump-houses, inserted steel-jaw traps on the shelves where the muskrats would sit, and then covered the holes. By noon, Willie and his brothers had set their entire string of traps, twenty-five in all. Because the weather was warm, muskrats, normally night feeders, were on the move in the daytime, so Willie patrolled the banks and open water with the family's .22 rifle.

In the next several days before the weather soured, Willie and kin trapped or shot more than one hundred muskrats, a good catch. After skinning and drying, the pelts would bring an average of 75 cents apiece, a much-needed boost after the long winter had drained the supply of food and money.

Willie's family lived mainly on salmon, sheefish, and cod, which they netted, and caribou, ducks, and rabbits, which they shot. But even this subsistence living could no longer be insulated from the cash economy. Fuel for outboard motors and ammunition had become necessities, and trapping was one of the few ways that Natives could earn the cash to buy them.

Work for cash was scarce in Willie's village, as it was in the other Eskimo settlements scattered along Alaska's coast on the Chukchi and Bering seas. Construction of government buildings created a few jobs during some summers. And laborers were needed at Kotzebue, the region's trading center, about 10 miles down the coast, to unload goods for the Eskimo villages on the Noatak and Kobuk rivers from ships onto barges headed upstream during the ice-free months. But the number of jobs fell far short of providing the necessary dollars. An anthropologist's survey of the Kotzebue area in 1954 counted only thirty permanent jobs in the town of 820 residents, paying an average of $250 a month. White traders controlled the stores and the few other commercial enterprises, pricing their products about 65 percent over Anchorage prices, which were already about 25 percent higher than Seattle's. Some called this gouging, but traders insisted that the markups were necessary for them to survive in isolated villages.

Then as now, about one sixth of Alaska's people were Native, descendants of the aborigines who had been there since prehistoric times. Slightly more than half of these were Eskimos, most of them settled on or moving along the coasts of the Bering and Chukchi seas in western Alaska. Eskimos are both maritime and land people, and are classified according to the two major language groups, the

Yupiks on the central and southern coasts, and the Inupiats in the northwest and north.

The Aleuts were the smallest Native group, constituting about 12 percent of the indigenous population. They lived, and still live, mainly on the Aleutian chain and on the Pribilof and Commander islands in the Bering Sea. Aleuts resemble Eskimos but have their own ethnic identity, including three distinct dialects. Since the volcanic islands they populate are treeless and windswept, Aleuts must live off the vast marine resources. Their skill as kayak hunters is renowned.

The remainder were Indians, who accounted for 36 percent of Natives. Known as Athabaskans, they inhabited the Boreal forests that characterize Alaska's interior. At least twenty different Athabaskan groups with eleven separate languages have been identified. Like those of the other Natives, their economies were based on a subsistence life-style—hunting, trapping, fishing, and gathering, plus whatever cash jobs they could find in the summer.

A late spring meant a crisis for the Native peoples. Meat spoiled even in the cool caches of the sandspit, temporary storage spaces usually dug into a creek bank, and some years fish runs under the ice failed to return. During extended winters, families shared dried salmon and whatever other staples they had, but cases of starvation and hunger-induced illness were common. So Willie and his family rejoiced at the break in the weather that enabled them to cash in on muskrats that spring. But far to the southeast, in Anchorage, Alaska's largest city, many were celebrating good fortune of another kind.

With the new Republican administration of Dwight Eisenhower in the White House, U.S. Secretary of the Interior Douglas McKay had started to release large areas of previously closed federal lands for oil exploration. At a bargain-basement rate of 25 cents an acre, Alaska oil leases immediately became a speculator's dream, even hotter than the rush for claims after the Klondike gold strike in the late 1890s. But those who'd joined the gold stampede could at least see pictures of nuggets actually dug along the Yukon. The new oil boom was rising on little more than an emotional high. At least 153 oil wells had been drilled since 1901, when the first recorded wildcatter showed up in Alaska, but not a single commercially productive well had been discovered. And the surplus of oil in the lower forty-eight made it hardly compelling for oil companies to explore in cold and costly Alaska.

Nevertheless, a deluge of oil lease applications swamped the federal Bureau of Land Management office in Anchorage. "A group of 17 Californians leased close to a quarter million acres," *The Anchorage Times* reported on June 23, 1953. "Anthony Zappa Jr. (no address given) posted a check for 99,022 acres," read another story. "A person who listed New York addresses put down a bank draft for a pile of oil leases four inches high," said another.

Outsiders showed the strongest initial interest, but on April 27, 1955, the paper reported that a local group of businessmen, including the town's leading banker, a store owner, a hotel operator, and the paper's own publisher, Robert B. Atwood, had filed for 15,360 acres of oil leases. They had set their sights on the Moose Range in the Kenai Peninsula, which extends just south of Anchorage into the Cook Inlet. For many, the entry of these normally conservative people legitimized the rumors of an impending oil strike. While Willie Hensley went on skinning his muskrats over the next few years, entrepreneurs in Anchorage were skinning oil lease customers. The most visible of the lot was one Locke Jacobs, Jr., a college dropout from Oregon who worked in an army-navy surplus store.

Because most of the vast acreages of public land in Alaska had never been surveyed, few people knew how to file a valid oil lease. With the help of a single geology textbook—which he says he never finished—and a bit of tutoring by a ship's captain he said he once worked for, Jacobs figured out how to mark off land from the few trustworthy meridians on early Alaska land maps. He also kept a record of every filing for gas and oil lands, as accurate as any kept by the Bureau of Land Management. Operating from the steps of the land office, Jacobs drew up leases at 10 cents an acre, sold maps at 25 cents a page, and brokered leases at a commission. As oil fever heated up, he sometimes made $650 a day by filing claims to Alaskan lands that neither he nor his clients had so much as seen.

Even by 1955, Willie Hensley had no inkling of these events, which would play so important a role in his future. The northwest Arctic where he lived was still virtually shut off from the world. No one in the village received a newspaper, and his family had no radio. Just surviving had occupied Willie's full attention for all his fourteen years. He never knew his father, a Russian-Jewish trader who had romanced his mother and then left her. Distressed and despondent, she had abandoned Willie before he could walk, leaving him in the care of Priscilla Hensley, her own adoptive mother.

Priscilla already had seven children of her own, but she took Willie in and adopted him, too. For the next several years, home for Willie was the one-room sod hut along a little stream that fed into the delta of the Noatak River. The hut was floored with small willow branches. Beds were boards laid on willow boughs and covered by caribou hides. A single window was made of beluga whale gut. The room was heated by a stove converted from a fifty-gallon oil drum. Fired with driftwood, it did the job even during winter days when the temperature hit thirty degrees below zero.

Generally, seven people—Priscilla; her husband, John; four of their children, and Willie—slept in that single room. The night's census depended on how many brothers, sisters, aunts, and uncles were down working or looking for work in Kotzebue, where the family had another one-room hut, this one built largely of driftwood. The toilet in either house was anywhere outdoors, but a "honey bucket" stood in the corner for use at night. In Itakatuq, it was Willie's job to empty it in the creek each morning.

Although his home lacked modern amenities, Willie knew it was an ideal location for a family that depended on fish, game, and driftwood. "The stream was so small that you could almost jump over it," Willie recalls. "But it was the source of a lot of fish. We used to set a net made of chicken wire at the mouth and we would get sacks of whitefish in the fall. Chum salmon would come in July and we could net sheefish and catch cod in the nearby Noatak River. Big flocks of ducks would fly by in the spring, ptarmigan and rabbits were plentiful, and caribou would pass nearby in the fall."

At the coast nearby, villagers gathered in teams to hunt seals and beluga whales. "Whales were an important source of our energy," Willie remembers. "We needed the blubber. Getting seal oil was also very important because we could store it to see us through the winter." The men did the shooting, but butchering was a family ritual. Women cut off the meat with *ulus,* sharp, crescent-shaped knives, separating the fat and emptying the intestines, a delicacy for the butchering party. The fat was rendered into oil and stored in sealskin bags. During the winter, fish caught through holes in the ice, frozen, and dipped in seal oil was a staple. It still is for subsistence families today.

Willie hauled water, gathered driftwood, and picked berries. When he was five years old, John died from eating spoiled *muktuk,* the white meat sliced from whale flipper. His eldest son, William, became the head of the clan, and Willie's responsibilities also increased. Another of Priscilla's natural sons taught him how to shoot, set traps, and net fish. That could easily have been the extent of Willie's education. "This is what we were trained to be—hunters and fishermen," Willie says. "To be anything beyond that, to be a professional, was basically beyond our imagination."

But one summer when Willie was seven years old, he made some friends on the beach and spent his days playing with them. "Then one day they were gone," Willie says. "I found out they had all gone off to school. I went back to the house where Priscilla was doing her chores. 'Hey,' I said, 'all my friends have gone off to school.' "

Priscilla did her best to squelch the subject. As far as she was concerned, children were supposed to stay outdoors during the day. "I guess I made a pest of myself," Willie says. "She finally put down her broom and said, 'Go ahead, go ahead, go to school!'" Willie walked to the one-room territorial school in Noorvik, not far from Itakatuq, and presented himself to Miss Logan, the white teacher who lived in an apartment in the rear.

While Eskimos do not live on reservations, the Bureau of Indian Affairs (BIA) ran the schools in the bush villages in those days. Teachers worked with the barest of tools and at times even dispensed medicine to ailing children between lessons. Willie thought first grade was fun, but he could stay only part of the year. The family needed him to help the hunt, catch fish, and store firewood for the winter.

The next fall, the BIA opened a larger school and brought in a new teacher, Miss Powell. When she asked students to seat themselves according to grade, Willie passed by the second graders and sat in the third-grade row, with his friends. A month later, when Miss Logan came in to compare notes with Miss Powell, she learned that Willie was in the third grade. "He doesn't belong there. He never went to second grade," Miss Logan said. "I don't know about that," Miss Powell replied. "But he is the smartest one in the class."

One hot summer day in Kotzebue when he was ten, Willie went to help Priscilla's daughter Jessie put down linoleum in a pool hall that Priscilla's husband operated, while the rest of the family went into town to look for work. While he was waiting for Jessie, a tall, slim white man approached him from the empty building next door.

"Do you want to work?" the man asked.

"Sure," Willie replied. "What do I do?"

Richard A. Miller was a Baptist minister newly arrived from Mississippi. He wanted to convert the empty building into a makeshift church. He put Willie to work cleaning out the building and, when that was over, gave him a job for the rest of the summer. He was to empty Miller's honey bucket each morning, at 25 cents a trip. "An easy job," Willie says. "His place was right on the water. All I had to do was take it to the beach and dump it in—just as everyone else in Kotzebue used to do."

When Willie was in the fifth grade, William's wife came down with tuberculosis shortly after giving birth to their first child, and Priscilla moved into their home, which was about 10 miles farther upriver, to help out. Willie went with her and pitched in by chopping wood and hauling water, but it meant he could no longer go to school. Miller, who had been impressed by Willie's quick mind and eagerness to learn, became concerned. Despite his own approaching marriage and the demands of his new congregation, Miller offered to let Willie stay with him in Kotzebue so he could stay in school.

"I found him a room with a bed," Miller recalls, smiling at the memory. "I believe that it was the first time he slept in a bed. Later, I took him on a trip to Fairbanks. There I showed him how to flush a toilet." More seriously, he continues, "There were two great qualities about him: He had super intelligence and he listened with that Eskimo quality of respecting elders."

Windows to the outside world began to open for Willie. He devoured *Newsweek* and *The Saturday Evening Post*. He discovered books and radio. And he studied hard, achieving eleventh-grade test scores by the time he reached eighth grade.

Miller and his new wife were planning their own family, but they held on to Willie, rejoicing at his progress and helping him come to grips with his background.

One Christmas he received a card signed, "Love, Annie." "Who the heck can that be?" Willie asked. Miller, who had learned about Willie's family from villagers, had to tell him he had an older sister he had never known named Annie. Another time, an Eskimo woman came to the door and introduced herself to Miller as Willie's natural mother. Miller had heard from parishioners that she was in town, and he knew she was looking for Willie. Miller brought the two together. "Willie, this is your mother," he said. Willie shook her hand. She was a stranger to him, and there was no joy in the reunion. After she returned to Fairbanks, Willie put her out of his mind and always spoke of Priscilla when he was asked about his mother.

After the eighth grade, qualified Native children were usually sent to Mount Edgecumbe, a boarding school maintained by the BIA in the town of Sitka on Baranof Island far down the southeast coast of the Gulf of Alaska. The courses were vocationally oriented. It was almost unheard-of for a Native to advance to college.

The Millers encouraged Willie to try for something better. "I don't know how deserved it was, but Edgecumbe had a poor reputation," Miller says. "The word was the girls came back pregnant and the boys returned drunkards."

Willie had saved some money working summers at Hanson's Trading Post in Kotzebue. The Millers pledged some of their own, and Miller's father in Mississippi, a deacon noted for his hellfire-and-brimstone sermons, also offered financial help, perhaps envisioning a promising convert.

All told, there was enough money to enroll Willie in Harrison-Chilhowee Baptist Academy, a boarding school just outside of Knoxville in eastern Tennessee, which charged only $400 a year. The appearance of an Eskimo in the heart of the segregated South puzzled his classmates as much as it did Willie. "I knew nothing about the Civil War and I was not prepared for the segregation I saw," Willie said. "But my classmates didn't know about Alaska either. I was considered simply a foreigner—like someone from Brazil or Chile—and we got along well."

Willie graduated in 1960, and made history for his village. He became the only member of his BIA school class—and all the classes he could remember before him—to make it to college. He attended the University of Alaska at Fairbanks for two years, then transferred to George Washington University in the District of Columbia. There began the emergence of Willie Hensley, Eskimo leader.

Living in the nation's capital, Willie began to form questions about the guaranteed equal rights he had studied in classes on the Constitution. During the summers, he worked as a typist in the BIA, where he met many Indian leaders. Talk inevitably turned to the Indians' fights to recover legal rights to the aboriginal lands stripped from them by a series of wars, treaties, and congressional edicts.

Alaska was the last great territory in the West where Native rights had not been contested, Willie told himself. With oil fever growing, what was going to prevent Alaska as well from being devoured by white people? Willie spent long hours at the library at the Department of the Interior studying the subject. He knew Eskimos in Alaska hadn't given serious thought to losing their land, and for good reason. They had lived on it for eons, and it was well established in their culture that land was held in common for the benefit of all. Besides, Eskimos had never been conquered in war as had the American Indians. And Native rights had been protected since the 1867 purchase treaty with Russia.

But by the time he finished his studies in Washington, Willie realized that the government was already selling and privatizing millions of acres of land in Alaska. "It took a while for me to get the connection," Willie says. Anyone who had the money and knew how to fill out an oil lease application—a category that virtually eliminated his own people—could get a piece of Alaska.

"We own it and we're losing it," Willie said to himself. He returned to Kotzebue after college. He was still grateful for a full summer's work handling the customers at Edith Bullock's busy tug and barge distribution center, but his naive trust in whites had gone. Once during that summer he turned to Ray Heinricks, the burly tug and barge operations manager, and said, half jokingly: "You know, Ray, one of these days we are going to run you guys out of our country." Ray responded in kind. "When you try to do that, Willie," he replied, "make sure you bring all your guns."

Guns were far from Willie's mind, however. He had a mission now, and he knew that the best way his people could fight back was with "white man's education." When summer ended, Willie headed back to the University of Alaska at Fairbanks to work on his master's degree—and to get ready to take on those who were selling his people's ancestral lands out from under them.

CHAPTER 2
The New Gold Rush

"MOVIE STARS BACK ALASKAN OIL HUNT
Bing Crosby, Laurel, Hardy Among Them
—Will Sink Exploratory Well"

THIS BANNER HEADLINE in *The Anchorage Times* on June 23, 1936, pierced the gloom that had set in over Alaska. The huge, untamed territory had always had a history of boom or bust, but for years now it had been all bust.

The big gold strikes at the turn of the century in the Klondike, Nome, and the Tanana River valley had created a mystique, but had little lasting impact on the economy. Most of the company gold mines that followed the prospectors were closed in World War II when the War Production Board deemed them nonessential to the war effort. Those that tried to start up later never succeeded in making gold mining a stable industry.

The Independence Mine at Hatcher Pass, one of Alaska's largest gold operations, shut down in 1958 and is now a museum. It blamed the fixed price of gold for its failure. While much gold remains in remote Alaska, the short work seasons and lack of roads in the mountainous terrain discourage new ventures, even though the price ceiling has been lifted. Analysts point out that South Africa in one year produces nearly as much gold as Alaska mined in eighty-five years.

For a time, it appeared that the enormous copper deposits found early in this century in the Wrangell Mountains might provide the territory of Alaska with the economic backbone it would need to qualify for statehood. The discovery spawned one of the world's greatest copper conglomerates, the Kennecott Copper Corporation, which merged the financial interests of J. P. Morgan with the mining interests of the Guggenheims. These giants of the "robber baron" era drilled mines into the mountainside, laid a railroad to transport the ore, and developed a small city—appropriately named Kennecott—in the midst of wilderness. The operation expanded rapidly and reaped enormous profits as prices for copper soared during World War I. Copper that cost Kennecott 4 cents a pound to produce was sold to the government at 35 cents a pound. By 1916, Kennecott had become the fifth-largest copper producer in the world, shipping out sixty-three million pounds annually. Even though prices dropped after World War I, Kennecott's presence continued to grow. By 1921, Alaska rose to third place among copper-producing states.

The bust came in a rush. Copper sank to a low of 5 cents a pound during the depth of the Depression in 1931, resulting in a $2 million net loss for the year at Kennecott. Then in the next year, floods washed out the firm's rail bridge across the Copper River. The corporation decided to close all Alaskan mines, ostensibly to wait until the price of copper rose above the cost of production. But with the richest deposits depleted, the mining conglomerate lost interest in Alaska and simply pulled out, abandoning its expensive railroad. It transferred its investments to Utah and Chile, where life was easier.

Kennecott quickly became a ghost town. Landslides and washed-out bridges made it inaccessible. By the end of the 1930s, paneless windows and roots thrusting through rotted floors were all that was left of Alaska's copper colossus. (The town is still inaccessible and continues to crumble to this day, despite periodic efforts to have it declared a historic site.)

In short, the news that an oil exploration expedition was heading for Alaska could not have come at a better time. The consortium of Hollywood explorers included not only rich movie stars—Boris Karloff and Spencer Tracy were also among them—but also well-known public figures and oilmen.

R. E. Havenstrite, a prominent Los Angeles businessman noted for his African safari adventures, was their spokesman. The group had invested upwards of $300,000, he said, and they were sending a freighter equipped with wildcat crews and equipment to drill an oil well at Iniskin Bay, on the west side of the Cook Inlet below Anchorage. "The Iniskin venture gives new evidence that the pioneering spirit that built California still lives," he said.

The Hollywood consortium would have fared better if it had used its stars and money instead to produce a movie on the perils and heartaches of oil exploration in Alaska. Land records show that the Iniskin Drilling Company did drill a well in 1939, but it came up dry. Havenstrite tried to salvage the natural gas that the company did find, but sparsely settled Alaska provided a poor market for it, and the meager discovery at Iniskin hardly justified starting up a transport system to the continental United States. The expedition became one more statistic in Alaska's long list of oil-seeking failures.

Eskimos, Indians, and Aleuts—the three major native populations—had lived on oil land for years, ignorant of its worth. Not so the white man. Oil in Alaska had tantalized ambitious individual explorers and some oil companies from the lower forty-eight even before the twentieth century, despite the intimidating obstacles.

Historians report that oil seepages were first recorded in Alaska in 1882 at Oil Bay, on the Iniskin Peninsula, the area that later attracted the Hollywood investors. A lone prospector referred to only as "Edelman" is said to have drilled two holes but abandoned his claim for unknown reasons. A group of drillers picked up the property in 1889 and hit oil at 700 feet. The well produced only 50 barrels a day, and when the drillers went down to 1,000 feet, they entered a water stratum, which effectively ended Alaska's first attempt at oil production.

Then, in 1896, a gold prospector named Tom White discovered oil when he accidentally fell into a pool covered with natural seepage at Katalla, about 110 miles southeast of Valdez. The story goes that he rinsed himself off and then filed the first oil claim. A firm named the Alaska Development Company drilled a well on the site. Oil spouted so spectacularly that the event merited a page-one story in *The New York Times* on September 18, 1902. "An immense oil gusher was struck at Cotella [sic] on the south Alaskan coast," the *Times* reported. "The gusher took everything away with it, rising nearly 200 feet before it could be capped . . . an important new industry is thus added to Alaska's resources."

Not quite. Four wells were drilled in Katalla and a one-street boomtown with six saloons sprang up. However, this promising beginning was marred by local feuding reminiscent of the battles over gold claims. According to an account in the *Alaska-Yukon Magazine* back then, rival drillers dumped "scrap iron and junk of all kinds" into competing wells. A firm named the Amalgamated Development Company bought out the holdings of the warring factions, but Katalla never fulfilled its high expectations. The company went into receivership, and several other firms tried to bring about profitable production until 1933, when a Christmas Eve fire destroyed the refinery and any hopes of establishing even a small oil industry in Alaska.

For sixty years after the first discovery attempt, American explorers punched more than a hundred holes in Alaska's arctic and subarctic earth probing for oil, but the total extracted—an estimated 153,000 barrels—would have barely filled one medium-sized tanker. They were all defeated by the same foes: gridlock ice, blinding snow, and fierce winds in the winter; floods, insects, and bears in the summer; equipment breakdowns, cave-ins, and remoteness from markets the year round.

Furthermore, with oil selling at only $1.00 to $1.50 a barrel for the first half of the century, the oil companies with the real financial muscle had no incentive to battle these harsh elements. It was simply better business to explore in more temperate and accessible places, such as California, Texas, and Oklahoma at home and Iran, Saudi Arabia, and Kuwait abroad.

Irene Ryan, geological engineer and frontier lady, was the first woman to fly solo in Alaska. "I came from an oil family in Texas, and you develop a feel about oil," she recalls, pert and articulate at eighty. "I saw oil discovered by dowsers who waved a willow branch over a map and by people who just had a hunch, but I had faith that, with emerging technology, major oil deposits could be found in Alaska in a scientific way. I flew over the state border to border in 1931–32 and marveled at the potential. I felt people should be looking for oil instead of gold. Everyone thought I was nuts."

Her husband, Pat, a contractor and her college sweetheart, dutifully accepted her causes, which led her to become the first woman to serve in both the territorial and state legislatures. And she tried hard to be the first person to drill a commercially productive oil well in Alaska.

"Oil and gas, like water, are constantly on the move," says Irene, who studied mining and petroleum at the University of New Mexico, where she earned her engineering degree. "They are continually trying to escape to the surface because of the force of gravity on the shale above it. Oil collects in what you call a structure, an anticline—call it a hill—in which a part then is capped by a tight seal. It sometimes is trapped in rock, or in shoestring sands found in the porous sections of ancient rivers, or in reefs where the ocean once was. Alaska has all these features, but the surface exposures that lead to promising structures underneath are harder to correlate because they are covered by glacial silt—glacial moraines."

In 1947, Irene was confident that she saw an opportunity for a major discovery. A mountain slide about 150 miles northeast of Anchorage in the Nelchina basin along the Glennallen Highway had exposed a 4-mile swath, uncovering beds of seashells. "I went up and talked to the locals and found some of them carried away seashells three feet in diameter," she says. "This turned out to be a cretaceous area, evidence that the ocean had been there at some point. And this was about the same time as the big oil discovery in Canada, at Leduc, outside of Edmonton. That was in a cretaceous field, too."

The similarities were exciting. Irene shared her discovery with a group of Anchorage businessmen who readily agreed to underwrite her further exploration with the aim of interesting an oil company. She spent the entire summer of 1948 mapping all of the cretaceous exposures. Then, confident that she had found characteristics of a promising structure, she wrote her report and took it to nearly all the major oil operators in the United States. She visited Richfield, Standard of California, Phillips Petroleum, and even famed wildcatter Glenn McCarthy and billionaire H. L. Hunt. "I almost got Richfield interested, but they had just started developing a new field in California," she says. "The other expression of interest was from H. L. Hunt's geologist."

The Hunt representative asked if she had a sufficient number of Alaskans to lease the land. She assured him that she did. After some months had passed, he called to say that while H. L. Hunt had his hands full in semiretirement, it was possible that his son, Nelson Bunker, then attending college, might want to undertake Alaska oil exploration as a project of his own. "Nelson turned it down, saying he was more interested in prospecting in eastern India," Irene says. "I am sure the real reason was that Nelson wanted no part of working among glaciers and bears, the common misconception of the state that still causes us problems today."

However, by the early 1950s, troubling events began to occur in the international oil fields—troubles that would force the companies to reconsider their forbidding image of Alaska. In 1951, the Prime Minister of Iran, Dr. Mohammed Mossadegh, who was hostile to all foreign controls in his country, nationalized Iran's entire oil industry, forcing the principal producer, British Petroleum, to withdraw after fifty years of operating in the country. Meanwhile, the big American oil firms watched uneasily as President Gamal Abdel Nasser of Egypt entered into arms deals with the Soviet Union, a splitting with the West that culminated in his attempt to close down the Suez Canal in 1956.

Suddenly, the hidden petroleum reserves of Alaska began to tempt the major oil firms. Robert B. Atwood, publisher of *The Anchorage Times* from 1935 to 1989, and still robust and square-shouldered at age eighty-four, recalls the influx: "We saw these oil people coming in here every spring, in their business suits and with their Harvard haircuts. They would hang their suits up in a closet and then go out of here in their rough clothes with pack trains and every other gear under the sun and they would head for the boondocks to do their fieldwork."

Oil companies' interest in Alaska grew to such a pitch that John Roderick, a Yale football star who had been lured north by the appeal of a free frontier, started a newsletter. In impeccable scholarly prose, the *Alaska Scouting Service* reported the names of oilmen coming into the state, where they were exploring, and what depths active drilling rigs were reaching.

Chevron appeared to be the first big player on the scene, arriving in 1953 and concentrating on the Kenai Peninsula. Soon after, Phillips Petroleum joined the search and obtained exploration rights on 1,063,034 acres in the Yakataga and Katalla districts, close to where the Hollywood consortium had given up. Standard Oil of California sent three geologists to Icy Bay, which lies between Cordova and Yakutat, in 1954. Humble Oil (now Exxon) conducted "extensive reconnaissance," according to the scouting service, in the Yakutat region in southeast Alaska, just north of Glacier Bay. Union Oil (now Unocal) began an aggressive hunt on the Kenai Peninsula.

Several local companies, with names such as Grubstake Inc. and Anchorage Development Company, also made regular news in Roderick's scouting reports. However, few people paid attention to the speculations that grew out of lunchtime conversations at the Anchorage Elks club. The regulars at the common table there included newspaper publisher Atwood; his father-in-law, Elmer S. Rasmuson, head of one of the two banks in town; a hotelman; and several store owners. A bartender referred to them as the Spit and Argue Club.

Atwood says they readily compared observations on the oilmen who came into town. As a journalist, Atwood made it a point to get information from the geological field teams when they returned. Others talked to them in the stores where they shopped and the hotels where they stayed. "Everything was so slow in those days," Atwood says. "They had to wait for steamboat travel, so they would be around for a week sometimes, sitting around a beer parlor, drinking beer. We'd visit and drink beer with them.

"Soon we noticed that a bunch of oil lease applications were being filed for people with California addresses." (*The Anchorage Times* faithfully published the names of people filing oil leases and ran brief descriptions of areas they sought to claim.) "We learned about oil leases from these people. We learned that oil companies don't want leases in their own name, but prefer somebody else to file, and then they will deal with them. In those days, an oil company was restricted to filing for up to a hundred thousand acres. They wanted to be sure that when their name was on a lease, it had oil on it. So they wanted someone out front leasing land that they would take when they wanted it. And they weren't interested in little pieces, but wanted a minimum of about sixty thousand acres before they'd talk."

Atwood, Rasmuson, and a dozen others, most of them luncheon-table regulars, decided to pool their funds and file oil leases based on the prevailing intelligence, which essentially was to guess which areas the oil technicians would recommend for drilling and then file on them, or next to them if someone else filed on them first. (Some rival businessmen of that era insist that the group's guesswork was considerably enhanced by eavesdropping on telephone conversations when the field men called their home offices from the Westward Hotel, a hotel owned by Wilbur Wester, a member of the pool.) "A group of Anchorage residents have joined the oil lease movement," *The Anchorage Times* reported on December 6, 1954. "They filed for leases on 65,280 acres of unsurveyed land north of Kachemak Bay on the Kenai Peninsula."

The group also included Locke Jacobs, Jr., the young army-navy store clerk who had been keeping daily track of leases filed at the land office. By now, Jacobs had become the foremost salesman of Alaska oil leases. At 25 cents an acre, they were a sure thing. They could be resold to oil companies at $1 an acre or assigned for a share of the royalties should a well come in.

"We had leases over at Glennallen, over at Beluga flats, over at the other side of the inlet, and down in the Kenai," Atwood says. The filing in the Kenai was in the National Moose Range, which begins about 35 miles southwest of Anchorage. The refuge was—and still is—a national treasure. Its diverse 1,730,000 acres contain scenic mountains, giant glaciers, thick forests, lowland lakes, famed salmon streams, and muskeg swamps. The range supports thousands of big-game animals—moose, bear, and caribou—and more than a hundred species of birds. Rainbow and lake trout lurk in its waters, and all five species of salmon migrate in its streams.

The area had been famed for the great herds of stone caribou that migrated there, but several forest fires in the early 1900s devastated their lichen ranges. Then miners and workers building the Alaska railroad nearly hunted out the weakened survivors. But the new growth in the burned areas made an ideal habitat for moose. And as the moose thrived, so did Kenai's fame as a big-game paradise.

Finally, the impact of the hunters it drew from all parts of the world became so intense that President Franklin D. Roosevelt was persuaded to sign an executive order in 1941 designating the Kenai range as a game refuge. The area was placed under the management of the U.S. Fish and Wildlife Service. Hunting would be controlled by provisions of the Alaska game laws, with approval of the U.S. secretary of the interior, and all lands would be protected from settlement. This led many to assume the refuge was off limits for mineral exploration.

By the end of 1955, oil companies and individuals had filed for more than 5 million acres, most of them in the south and southeast, since the northern tier was off limits by Defense Department edict. Still, not a single commercially productive oil well had yet been drilled anywhere in the territory.

The high hopes generated by the sight of oil company geologists, seismic crews, and oil rigs swarming over Alaska began to dim as news stories and scouting reports indicated that the sites were producing only dry wells. Seeing the oil boom starting to sputter, the Anchorage luncheon-club investors made a brash decision. Instead of holding out to sell leases at a profit, they agreed to offer their leases free to any oil company that would drill a well on their land within two years. What did they have to lose? Until it showed the means to support itself as a state, Alaska would remain a ward ruled from Washington and dependent on federal doles. And as long as that continued, the businessmen who gambled on investing in the land of the future faced a bleak return. Alaskans would continue to be second-class citizens.

Atwood had been among the first to recognize this. He had been campaigning for statehood almost from the time he came to Alaska in 1935 to run the *Times*. To boost his cause, he stressed the state's strategic military location. He criticized the mining and fishing interests that opposed statehood because they did not want to pay state taxes to build roads and schools. His paper became the voice of statehood supporters. And his energetic wife, Evangeline, organized a drive for statehood across the territory.

The Japanese attacked Attu in the Aleutian Islands in 1942, the only American territory on the continent invaded by the enemy in World War II. That focused national attention on Alaska's strategic military value. The Cold War with Russia following the armistice further reinforced Atwood's argument. However, the message wasn't impressing Washington. President Eisenhower, as late as 1954, stated that complete control by a central government was the best way to meet any security threats from across the border.

Statehood bills had been introduced in every session of Congress during the late 1940s and the early 1950s, but U.S. House Speaker Sam Rayburn throttled them before they could reach the floor. The southern bloc in Congress feared the addition of two senators from a new northern state, who they believed would tip the balance against their effort to hold back by filibuster the flood of civil rights legislation then building up in Congress.

Traveling at his own expense, Atwood lobbied hard for statehood support among his newspaper colleagues at American Newspaper Publishers Association conventions, urging them to apply editorial heat on Congress. And Alaskans had voted overwhelmingly for statehood in 1946. But a decade later, Congress was still stonewalling with the same theme: You can't show us you have an economic base to support statehood. It was true. The salmon industry was slumping, endangered by overfishing by the Seattle interests that controlled the canneries. Mining had seen its best days, and timbering needed heavy federal subsidies to survive.

Atwood could also see that Anchorage, the closest thing to a metropolis in the state, was struggling. It had valiantly tried to modernize, but a large part of the city was still mired in the frontier. Many streets were unpaved. Housing was meager, shops and restaurants few. Saloons stayed open around the clock, burlesque dancers cadged drinks from bleary-eyed patrons, and all-night poker and dice games attracted unsavory characters who provided a steady stream of police news—and frequent obituaries. Downtown businessmen could only eke out a living, competing for the limited revenue brought in by military bases. One sharp federal budget cut in the military could put stores and banks out of business.

On June 1, 1955, Roderick's scouting service reported an unusual item. It said the Atwood group had sent Wilbur Wester, the hotelman, on a trip to Southern California to interest oil companies in Kenai lease holdings. His arms loaded with maps, Wester had been appointed to carry out the group's bold decision to offer the leases free to any company willing to drill a well on their property.

A weary and discouraged Wester returned without a nibble.

Next, Locke Jacobs, Jr., took his own sizable lease agenda beyond California, approaching nearly every major oil company with the story of Alaska's potential. Not only was no one interested, he found, but he was practically thrown out of some offices.

By now, critics were openly jeering at efforts to make Alaska an oil state. One officer at the First National Bank of Alaska, the rival to Rasmuson's National Bank of Alaska, was reportedly advising customers not to fall for the oil leasing game. "You got a better chance of getting your money back by throwing it into the Cook Inlet. At least there the tide comes in twice a day," he was quoted as saying.

"We pleaded with the oil companies to at least drill a hole," Atwood says. "We tried to impress on them that what motivated our group was not the fast buck—we were interested in developing Alaska." But the big oil companies simply weren't accustomed to listening to uninvited visitors who crossed their thresholds bearing maps. The Anchorage group simply aimed too high. In the end it was a much smaller player, Richfield, that decided on its own to take on Alaska.

Richfield was founded in 1911, when Kellogg Oil and the Los Angeles Oil and Refining Company jointly financed a plant to top more Santa Fe railroad crude from the Olinda oil field in Orange County, California. They built the plant at Richfield station, a railroad siding about 3 miles from Olinda. The stop included a grocery store and a railroad watering facility. When the two companies decided to merge in 1915, neither wanted to lose identity. But both liked the sound of Richfield, so they compromised by adopting the name of the siding.

The humble birth of Richfield was capitalized with $70,206. Besides the topping plant, its assets consisted of two small refineries in Los Angeles, a tank wagon, and a team of horses. But the company grew in the seller's market of World War I and through the 1920s, as trucks replaced wagons and automobiles replaced horse-drawn carriages. The company blundered, however, with the purchase of Pan American Western Petroleum for $7.5 million in cash in 1928, a white elephant that accelerated Richfield's crash when gasoline prices plunged to 9 cents a gallon a few years later. In Richfield's operations, that didn't even cover costs.

In the early 1950s, while oil speculation in Alaska was reaching a fever pitch, Richfield was scarred by bankruptcy and a tortuous reorganization following four straight years of declining earnings. It had no international presence, and with troubles in the Mideast driving up the cost of crude, as the company's new president, Charles S. Jones, would later note in his autobiography, it was imperative to develop its own American reserves.

A little-publicized ruling by Eisenhower's secretary of the interior, Douglas McKay, undoubtedly helped boost the company's interest in Alaska. In 1954, after leasing applications had been briefly halted pending a study, McKay declared that the multiple use of withdrawn federal lands was in the best public interest, clearing the way for exploration on the Kenai Peninsula's Moose Range. The oil industry had, of course, been Ike's major financial supporter in the 1952 election. And the opening of the Kenai was only the beginning. McKay seemed hell-bent on paying off political debts. In fact, he relaxed the rules to favor private exploitation of federal lands at a pace that would eventually lead to a congressional investigation.

In 1954, Frank Tolman, a geologist in Richfield's Bakersfield office, made a study of U.S. Geological Survey maps of Alaska and found an impressive topographical high on the Kenai Peninsula. Upon his recommendation, the company decided to lease the land. A landman, George Shepphird, went up to Alaska and leased 70,000 acres in the area.

However, even with that commitment, Richfield was still the puniest horse in the great Alaska oil sweepstakes. According to the scouting service, Phillips Petroleum by then controlled more than 1 million acres under lease, Standard of California had 439,202 acres, Anchorage Independent had 405,000 acres, and Shell had 199,000 acres. Union and Marathon, working together to explore the Kenai, controlled 254,440 acres, more than three times the size of Richfield's little plot.

But early in March 1955, Mason Hill, Richfield's chief geologist, called geologist William Bishop into his office. "You're going to Alaska," he said. "I don't know what you need to do to find oil up there, but you figure it out."

CHAPTER 3
The First Big Strike

William C. Bishop arrived in Anchorage on March 15, 1955. Before the night was over, he could sympathize with every wildcatter who had come before him in search of oil and been defeated. A 4-foot snowfall stranded him in the Westward Hotel in the heart of the city and made any site exploration impossible. After a week, he returned to the airport and caught the first flight home.

Returning in May, a more agreeable month in the subarctic, Bishop headed for the birch, aspen, and spruce lowland forests near the Swanson River in the Moose Range. The snow was mostly gone, but he found himself slapping mosquitoes, shooing away moose, and looking over his shoulder for bears. His overriding concern, though, was how to fulfill the mandate he'd been given.

Gentle, clean-shaven, almost shy, the thirty-six-year-old Bishop did not look like a rugged adventurer ready to take on Alaska. He had grown up on a ranch near the little town of Hartley in the Texas panhandle, where hot, choking dust was a way of life. When World War II came, he joined the Marines because he wanted to be a pilot. To his dismay, he found out the Navy trained all the Marine pilots. He ended up in a mortar platoon and spent the next two years on steamy jungle beachheads at Bougainville, Guam, and Iwo Jima as the United States took the South Pacific island by island.

He was hired by Richfield as a geologist before he had completed his master's thesis at UCLA. His toughest assignment with the company until Alaska had been a geological expedition to Dhofar, a remote region in Oman. His six-man field party had to cope with 120-degree desert days, feuding tribesmen, and a sullen sultan who was to direct them but didn't even know where his own domain ended. Nevertheless, the American team managed to produce the first geological survey of that uninhabited area.

Now Bishop faced an even harsher challenge. And there was a new aspect to it as well. An environmental awareness was dawning in the United States. Congress was becoming uneasy about Secretary McKay's permissive attitude toward oil exploration in federal refuges. Alaska still had no organized environmental watchdog group, but by 1955 national conservation organizations had definite doubts about McKay's commitment to protect the natural resources of the nation. Ernest Swift, executive director of the National Wildlife Federation and a former U.S. Fish and Wildlife Service officer, called McKay's leasing order "the closest thing to an outright giveaway yet perpetrated on the wildlife refuge system."

Before Richfield could get a clearance to work on the Moose Range, it had to pass muster with Congress. In 1955, the United States Senate Committee on Interior and Insular Affairs held hearings on Richfield's application. Charles S. Jones won approval after satisfying the committee that his company would follow government guidelines and clear all development plans with the Fish and Wildlife Service.

David L. Spencer, supervisor of Alaska refuges for the Fish and Wildlife Service, recalls drafting the regulations for oil drilling. "We tried to enforce a minimum intrusion," he says. "They were to cross the creeks in certain ways, no debris in the streams, and we banned dynamite for seismic studies. We had a separate set of regulations for cleanup afterwards . . . they couldn't leave gouged areas and they had to revegetate anything they tore up."

Bishop, used to the freedom of exploring in desolate areas, had to adjust his style. No longer could he use the explosive nitramine for seismic shots; instead, he had to resort to hydrophone soundings along the lakeshores. This returned weaker signals, but it did the job well enough to convince Richfield of the potential of a topographical high. "The streams kind of went around the high," Bishop says. "We saw some oil seeps on the west side of Cook Inlet. Then we did some geophysical work with the helicopter and set off several more soundings."

Based on seismic explorations from only thirty-three shot points—a skimpy number for a drilling decision—Richfield was confident the readings showed a large structure capable of trapping oil, though there was no way short of drilling to be sure. Bishop marked the location, next to a red-bark hemlock—the only one in the area.

However, after studying the map, Bishop saw an unexpected problem. The possible oil-bearing structure underneath extended more north-south and into property that Richfield had not leased. The company agreed that it should control the adjoining property, which is how Bishop finally met Locke Jacobs. The land was on the Kenai leases Jacobs had acquired for the Elks Club luncheon pool. In a deal struck in November 1955, the group was happy to unload 60,000 adjoining acres for 25 cents an acre and an override should Richfield strike oil. "We told them we only wanted some of the land, but they insisted that we take the whole block," Bishop says.

Early the next year, at a meeting in its Los Angeles office, Richfield had a critical decision to make. Venezuela was opening leases for oil exploration after a twenty-year closure. And there were the promising leases in Alaska. Some at Richfield wanted to invest all of the firm's limited exploration money in Venezuela, pointing out that drilling in Alaska would cost twice as much and that many other companies drilling up there had only a long string of dry holes to show for it. Bishop argued for Alaska. Richfield's president, Charles S. Jones, made the decision. On to Alaska.

The Swanson River site was deep in the forest, 23 miles from the nearest road, so the wildcatters would need a way to get in. Bishop laid out the road in keeping with Richfield's low-budget style. Flying at treetop level in a Piper Cub with a case of toilet paper at his side, he unfurled a paper trail for the bulldozers to follow from the highway to the red-bark hemlock. Then he returned to the site, twisted a bootheel in the snow, and said, "Drill there."

Drilling started in April 1957, but Bishop could not stay for the results. He took off for Italy, his next exploration site, but first he left orders. A colleague, geologist Ray Arnett, was chosen to "sit on the well," a job that entails checking core samples as drilling proceeds. And he left Arnett, who would later serve as assistant secretary of the interior under James Watt, with a secret code and orders to telegraph him when anything definitive developed.

On July 15, 1957, Bishop was back from Italy and fast asleep in his home in Inglewood, California, when a wire from Arnett arrived at three o'clock in the morning. "We are cutting wood," it read. Bishop scrambled for his suitcase. "Hope you have enough wood to make a short table," he wired back. That told Arnett to get a core large enough for a test.

"I arrived the next day in time to test it," Bishop said. When the final core was pulled, a gusher of oil exploded. The first well in Alaska capable of producing oil in commercial quantities had been discovered. The high pressure and volume made it necessary to stop the test, but rough calculations put the total output at 900 barrels a day. It was the kind of bonanza claim holders in Alaska had been praying for.

"Richfield Hits Oil," announced 3-inch headlines in *The Anchorage Times*. "Feverish excitement has spread through the area. . . . A lease filing rush starts," blared stories on the front page. "The best advice for Alaskans today is that of a San Francisco cable car conductor to his passengers: 'Hang on, we're going around a curve,' " Atwood wrote in his editorial. "Alaska is turning a sharp curve and is starting down a new road of development such as never has been seen before."

The well, which showed oil at 11,000 feet, was soon producing the 900 barrels a day it had promised. In the world of oil exploration, the discovery was extraordinary. Not only did it occur on the very first drilling attempted in Alaska by Richfield, but Bishop himself admits his calculations hit the mark by the very slimmest of margins. The drill had barely caught the extreme northern end of the oil field. Had Bishop twisted his boot 100 yards farther away from the red-bark hemlock in the other direction, Swanson River would have become only a trifling land office statistic—the 166th consecutive nonproductive oil well drilled in Alaska.

The First Big Strike

Locke Jacobs heard the news of the discovery in midafternoon while on duty at the army-navy surplus store. He rushed out the rear door, and this time he wasn't coming back. Jacobs dashed down Fourth Avenue, the one fully paved street in Anchorage, and spent the night at his post on the front steps of the federal building preparing for the rush on the land office. By morning, with the assistance of a newly hired secretary, he was ready to accommodate the mob straining to take out oil leases. He estimated he earned $1,000 an hour for every hour he could stay awake.

Jack Walker, an Anchorage barber who also had learned how to file oil leases, hung a "Closed" sign on the door of his shop next to the Pioneer Bar on Fourth Avenue. He never returned to cut another head of hair.

"I practically camped at the land office after that," the frail, genial Walker says. "Every oil company in the United States was sending somebody up there. I sold the leases to ordinary people hoping for a killing. I sold leases to people fronting for Standard. They called them 'nominees.' They would come to me. All I had to do was beat them to the land office.

"It was first come, first served. Later, when so many people were filing that they had to hold drawings, I rounded up a lot of people to put their name in the hat. If they won, I'd get a little cash and maybe a percentage of any oil royalties."

Walker says he never stopped to count the money he made, but he bought an office building downtown on K Street, which he rented to Sohio and Amoco. He also bought a big stretch of waterfront property on Nancy Lake, on the way to Denali, and built a cabin on it. "And I guess I also just plain blew a lot of money," he adds.

The Bureau of Land Management office in downtown Anchorage became a madhouse. The news pulled miners from their mines and housewives from their kitchens. Filings poured in so fast that Virgil Seiser, the embattled land office manager, was castigated by impatient throngs because his office staff could not keep up. The backlog of unrecorded leases made it difficult to determine which lands were still available—an Alaskan version of chaos in the stock exchange.

Seiser was soon accused of favoritism. When Locke Jacobs, whose own records of land filings always seemed to be more current than anyone else's, was permitted to set up a photocopying machine inside the federal land office, suspicions of a dubious partnership multiplied. Jacobs claimed that he shared the copier with Seiser's swamped clerks at no charge and that this technological assist improved the government's output. However, when only Jacobs had a supply of filing forms after the land office ran out, the clamor for Seiser's scalp grew louder.

Many of those who tried to get in on Alaska's leasing boom aggressively vied with Jacobs and Walker for hints of inside information. Some set up scouting services and sold their reports much the way tip sheets are peddled around racetracks. And one enterprising scout, dressed as a gold prospector, had himself dropped out of a helicopter, dog team and all, so he could spy on Mobil drilling operations at the tip of Yakutat.

"We all tried to figure out where the big companies were filing, then we would file next to them," Walker says. "Nobody knew any geology at that time. This was a learning process for everyone, including poor old Virgil, who had to deal with the technical questions at the land office."

It was also a con man's field day. "A lot of shady people came up here to rip off people in the States who didn't know better," Walker says. "I know some people were told they were buying forty acres of Alaska oil land when the guys taking their money were actually only leasing forty acres and no one had any idea whether it had oil."

Seiser, trying to stem the most reckless land claiming, refused to grant leases to a group of Seattle investors who wanted to lease on the Malaspina Glacier, which fronts Yakutat Bay. When Atwood editorially defended Seiser's decision and Jacobs made derogatory remarks in public about people who would drill on a moving glacier, the rejected group's lawyer, Edgar Paul Boyko, filed a $1.5 million libel suit against all three. The suit was dismissed before it ever came to trial, but the Seattle interests had other ways to get back at Seiser. They managed to persuade their congressman, Thomas M. Pelly (R-Wash.), to investigate "reports of favoritism, claim jumping, and lease sharking in Alaska."

"We may have a national scandal in the making rivaling any in our past history," Pelly announced. "If some clerk in the BLM land office decides an application is not correct in every detail, the news and official number of disputed units travel at an amazing rate of speed . . . lease sharks with connections in every land office in Alaska and through into Washington, D.C., are operating in the territory in an effort to take advantage of technical faults in oil lease applications."

Seiser maintained he had denied the leases because he believed no one should drill on a moving glacier. Nevertheless, he was transferred to Portland, Oregon, while the Department of the Interior investigated Pelly's charges. Nothing came of the investigation, but Seiser's reputation in Alaska had been blackened. He served out the remainder of his career in Oregon and retired with commendations from the Department of the Interior, a belated attempt to make up for the abuse he'd taken.

In the year since the Richfield discovery, leases for more than 33 million acres of federal land were issued in Alaska. Natives, however, who had lived on the land for more centuries than is precisely known, were not among those who held them. As Willie Hensley says, they hadn't acquired white men's prowess in such procedures. Moreover, as the hordes rushed in to claim a piece of Alaska, no one considered whether the leasing impinged on Native rights. The Kenai Peninsula had been settled by the Kenaitze Indians long before the arrival of whites. Ancient Indian village sites with their moss-thatched log huts were still visible in parts of the peninsula.

And many Native fishermen and subsistence families lived on both sides of the Cook Inlet, though their distinct cultures had been blurred by the influx of commercial fishermen, homesteaders, and service people working in hunting and fishing lodges. But the oil rush in the Kenai raced on without anyone raising the question of aboriginal rights. It came to a dead stop, however, over the issue of moose rights.

Late in 1957, Richfield drilled a confirmation well at Swanson River. It promised to be a larger producer than the first, but the rejoicing was quickly stilled by events developing in Washington. By now, the U.S. secretary of the interior, now dubbed "Giveaway McKay," had gone too far. Led by the National Wildlife Federation and the Wildlife Management Institute, conservationists called on Congress to rein in the secretary, who had "thrown open" 250,000 acres of wildlife refuge to oil and gas drilling. They cited one lease, for 12,000 acres in the Lacassine refuge in Louisiana, that had been granted without public bidding at "a ridiculously low fee of 50 cents an acre and 12 and 1/2 cents oil royalties."

The House Committee on Merchant Marine and Fisheries, chaired by Herbert Bonner (D-N.C.), had convened two months of hearings on the complaints. Bonner harshly criticized the Interior Department for leasing refuge lands for oil and gas exploitation so freely. Henceforth, he told McKay, all future requests for adverse use of refuge lands must be referred to the congressional committee for approval.

McKay resigned shortly thereafter, in 1956, to run for a U.S. Senate seat in Oregon, which he lost to Wayne Morse. President Eisenhower appointed Fred A. Seaton, a favorite assistant, to take over the Interior Department. Seaton, a Nebraskan who owned a string of newspapers in four western states, was regarded as a political pro, adept at pulling White House chestnuts out of the fire. He had helped shape administration strategy on farm and defense matters, and is said to have masterminded Vice-President Nixon's famed "Checkers" speech to the nation.

On taking office, Seaton declared a moratorium on further oil and gas leasing in national wildlife refuges and ranges. Then he closed down all oil operations in progress on those federal lands, pending studies to determine their impact, including the effect upon the moose in the Kenai.

The order devastated Alaskans, and they rose up almost in a body to protest the closure of their one potential year-round industry. Atwood unleashed his full editorial fire and even lobbied Seaton, a fellow member of the American Newspaper Publishers Association. (Atwood's actions were a clear conflict of interest for an editor by today's standards, but editorial advocacy mixed with personal confrontation was a brand of frontier journalism freely practiced by the Alaska newspapermen who battled for the birth of their state much as their counterparts in colonial times had fought for the birth of the nation.)

Help was forthcoming from other quarters as well. A report by Alaska Fish and Game, guardian of the state's wild animal resources, did much to blunt the fears of conservationists. It maintained that moose would not be bothered by oil rigs and might even welcome the browse from newly felled trees. But the most persuasive voice of all came from Ernest Gruening. President Franklin Roosevelt had sent Gruening to Alaska as territorial governor in 1939. The president could not have picked a more unlikely candidate. A Harvard man, an urbane Bostonian, and a fierce New Dealer, Gruening seemingly had nothing in common with the rough Alaskans of his era. He didn't even hunt or fish.

But a lifelong idealism bound Gruening to Alaska. He despised colonialism, and it did not take him long to feel the neglect and discrimination Alaska suffered in the custody of a distant federal government. He knew that life would never improve for Alaskans without statehood, and he continued to champion that even after the Eisenhower administration replaced him as governor of the territory.

In 1956, Gruening, now Alaska's provisional senator, joined the battle to reopen the Kenai Moose Range. He was so convinced that Alaska needed an oil industry to achieve statehood that he risked his standing as a dedicated and longtime conservationist. On December 9, 1957, the Department of the Interior opened hearings on Seaton's closure. Gruening was the state's star witness. Likening the proposed regulations to restrict oil and gas leasing on federal wildlife lands to "a prospectus of an obstacle race," Gruening said: "Having served eighteen years in federal government, I can recognize in the language the plasm of protracted procrastination. Or at least its possibility." He pleaded that Alaska be permitted to develop its emerging oil industry without being choked by bureaucracy.

Shortly afterward, Seaton invited eight prominent oil industry leaders and eight officers of the nation's leading conservation groups to a conference at his office. He asked the conservationists to sit on one side of the table. "Mr. Secretary, how do you tell the difference?" Richfield's president, Charles S. Jones, asked. After a ripple of laughter, according to Jones's account, the group got down to specifics in a constructive mood. As an outgrowth of the conference, the American Petroleum Institute set up a staff to serve as liaison with the conservation groups, its purpose to head off environmental disputes before problems arose.

In midsummer 1958, Seaton reopened the western half of the Kenai Moose Range to oil development. He used his executive powers to free Richfield and other oil companies to continue developing on the Cook Inlet side of the peninsula. Some contend that President Eisenhower, who benefited handsomely from oil industry campaign funds and contributions to his farm at Gettysburg, Pennsylvania, helped Seaton make up his mind.

But the secretary also recognized the environmental mood growing in the nation. Not only were the conservation groups becoming more vocal, but consciences were stirring inside government itself. In-house reports detailed dissension over Interior Department policies from the new breed of federal workers who had come on board to protect the nation's resources.

In Alaska, the Fish and Wildlife Service, swinging with the national mood, had protested McKay's permissiveness long and loud. It appealed to Seaton to eliminate from the new order thirteen ranges and refuges that contained rare species. Respecting those concerns, Seaton barred mineral exploration entirely in the eastern half of the Moose Range. Today, this is the most popular destination for tourists, who canoe and raft the turquoise Kenai River, fish the spectacular salmon runs, hike mountain trails leading to incomparable views, and camp along lakes teeming with trout.

Yet Seaton's sensible compromise hardly gave Richfield clear title to proceed. No sooner had the ban been lifted in 1958 than new problems developed. A group consisting of rival Alaska businessmen filed claims for nearly all of the land that the Atwood Elks club pool had leased and sold to Richfield three years previously, a practice known as "top-filing." Sometimes top-filers did win, particularly if they could show a defect in the original leases. Sometimes rivals bought them off, just to be rid of a legal nuisance. And sometimes, as in gold rush days, top-filers were shot. But most of all, top-filers were inevitable.

"Discover a major new oil field and litigious-minded people are legion," Richfield's Charles S. Jones wrote in his autobiography. "That has been our experience for over 40 years . . . and Alaska proved no different." The new claimants—who included Walter Hickel, later governor of Alaska and Richard Nixon's secretary of the interior—argued that since they were the first to file after the Moose Range had been divided into leasing and nonleasing areas, they had the first valid leases.

On April 10, 1959, the Bureau of Land Management office in Alaska rejected the claims and voided the leases of the rival group. The top-filers then placed their case in the hands of an aggressive Anchorage attorney, James Tallman. They pledged to take their claim to the Interior Department in Washington and to the federal courts, if necessary.

While the appeal languished in Washington, the stakes in the outcome suddenly skyrocketed. On March 15, 1960, Richfield, now joined as an operations partner by better-heeled Standard Oil of California, struck oil on an even larger field on the Kenai Peninsula. The new well, called Soldotna Creek Unit 3, was drilled on the leases acquired by the Atwood group. The royalties promised to return millions of dollars to the fourteen investors who held the original leases—or to their challengers, should they prevail.

The Interior Department in Washington upheld the ruling of its Alaska officers. The rivals then sued the Department of the Interior in federal district court, which dismissed the suit outright. However, on appeal, a circuit court handed the litigants a stunning upset. The case progressed to the Supreme Court, and Richfield and Standard joined the government in its defense, bringing in two of the most respected legal minds in the profession, Abe Fortas, who later became a Supreme Court justice, and Clark Clifford, who later became U.S. secretary of defense. The two filed friend-of-the-court briefs in support of the Department of the Interior.

On March 1, 1965, the seven-year battle over the oil leases on the Moose Range finally ended. The U.S. Supreme Court ruled 7–0 in favor of the Interior Department's right to support the original leases, with Justices Douglas and Harlan not participating. Chief Justice Earl Warren, writing the decision, stated that the secretary of the interior's interpretation of the original orders as not barring leases was "reasonable, consistently applied, and a reported matter of public record." Since leases had been developed in reliance upon it, the courts must give it credence.

In 1992, oil was still flowing in commercial quantities from the wells on the Kenai. Based on the standard 5 percent royalty which the Atwood group retained in its deal with Richfield, it is estimated that the fourteen investors had earned about $3 million apiece by 1990. Atwood, Jacobs, and several others bought many leases that did not produce oil, but some of the others are believed to have laid out less than $1,000.

More than thirty years later, various versions are still resurrected of how Bob Atwood and his companions received the information that led them to file leases on the first major oil-producing field in Alaska. Howard Weaver, as editor of the now defunct newspaper *Alaska Advocate,* implied that Atwood had insider information from high political levels. John D. Hanrahan and Peter Gruenstein repeated a similar rumor in their book, *Lost Frontier.*

In 1990, Weaver, now editor of the *Anchorage Daily News*, which was enmeshed in a bitter circulation war with the *Times*, sent a top political reporter on a six-month leave to retrace the oil discovery saga. The eight-part series turned up interesting research on the era, but failed to substantiate an insider plot. In fact, the series drew bitter reader responses, most of them complaining that the *Daily News* was repeating the slurs of "sore losers" in an attempt to damage the reputations of Alaska's heroes.

Besides, others had reaped the benefits of oil discovery as well. After Richfield struck oil, waves of drilling crews representing just about every major oil company headed north. In 1959, Union Oil and Marathon, exploring farther south on the Kenai Peninsula, discovered the first commercially productive natural gas well in Alaska. Eventually, four separate oil fields were found, and wildcatters even tamed the treacherous tides of the Cook Inlet to erect fifteen oil-producing stations offshore. The peninsula area remains an important source of energy to this day.

Thanks to the foresight of Irene Ryan, Alaskans were ready to collect the benefits when oil started to flow. She kept her colleagues in the territorial legislature in extended session until they passed an oil tax during the oil lease rush of 1955. Sensing someone would soon strike oil, she insisted upon and won a levy of 1 percent on the gross value at the well of all oil and gas produced within Alaska. This produced $314,000 for the state during the first year that oil flowed. It grew to $84 million annually in the next six years—the forerunner of Alaska oil revenues that would swell to intoxicating dimensions a few decades later.

Moreover, the oil strike at the Moose Range won Alaska credibility and respect, as well as giving the destitute territory its own revenue. The emergence of an oil industry finally convinced Congress that Alaska was worthy of statehood, which it achieved in 1959—less than two years after the Richfield discovery but twenty years after the Atwoods first began championing the cause. Many feel Alaska carried Hawaii into the United States as well.

And the fate of the moose on the Moose Range? Swanson River in the early 1990s had one of the highest densities of moose in Alaska, some 3.7 moose per square kilometer, compared to an average of 2.4 in interior Alaska, according to Chuck Schwartz, director of moose research at the Kenai National Wildlife Refuge, as the federal preserve is called today. While Schwartz finds that oil development has been environmentally compatible on the peninsula so far, he credits nature—specifically, regrowth from still another forest fire, this one in 1967—for keeping moose in such healthy numbers on the Kenai oil patch.

A special victory belongs to Bill Bishop, Richfield's modest geologist. He is not even mentioned by name in Charles Jones's book, *From the Rio Grande to the Arctic.* But Alaskans bronzed Bishop's boots and placed them on permanent display in the Anchorage Museum of History and Art.

CHAPTER 4
Race to the Arctic

When Boyd Brown, ace Caterpillar skinner for the Frontier Company, an oil contractor, returned to his Kenai Peninsula home in Soldotna one day in mid-March 1964, he had put in twelve hours of moving earth at the Drift River oil terminal. The last thing he needed was that phone call from Fairbanks. His boss, John C. "Tennessee" Miller, was on the line, and he was mad. "They goddamn near drowned in a beaver pond, Cat and all," Miller shouted. "And all they are doing is tearing up equipment. I need you up here." Things clearly were not going well for Miller's Cat train, which was trying to blaze the first overland route to the frozen North Slope on the edge of the Arctic Ocean, at the very top of coldest Alaska.

Oil explorations had shifted to the northern tier of Alaska with relaxation of federal controls in the 1960s. The federal government had treated the state as a war zone during World War II, even restricting travel and censoring mail, and had continued to keep the north off limits to all except the military right into the Cold War. But when the discovery at Swanson River intensified pressures for the development of an oil industry in all of Alaska, the feds lifted the ban on oil and gas exploration north of the Brooks Range as well.

However, the oil companies promptly encountered trouble moving the tons of equipment required for seismic explorations and the still heavier rigs needed to drill test sites. The slope is accessible by sea at Point Barrow for only three months of the year, and Barrow is hundreds of miles from the unexplored areas above the Brooks Range. And no one dared try to move anything across the uncharted, roadless terrain. Finally, hampered by the costly and time-consuming practice of flying in dismantled exploration equipment that then had to be reassembled, often in subzero weather, Richfield Oil promised Tennessee Miller good money if he could deliver Cats and sleds by land.

Miller put together a train of three D7 Caterpillars hauling a bunkhouse and two giant sleds filled with fuel, food, and equipment. He picked a four-man crew and sent them north with promises that an airplane would help them navigate and drop supplies along the way. No one except Miller—and he was a transplanted southerner—really believed that men could drive bulldozers into this largely mountainous wilderness during turbulent March weather without grave risk to themselves and their machines. It was 472 miles from Fairbanks, where good roads ended, to Sagwon, where seismic crews had established a staging area and waited for equipment.

After battling fierce weather for eighteen days and covering only 60 miles, the men on Miller's Cat train were near rebellion. The crew had expected the forty-degrees-below-zero temperatures, which, they were warned, instantly numb any exposed flesh. But they could not cope with the violent, unpredictable gales.

Without warning, blasts whipped snow into whiteouts—squalls so blinding that they made it impossible to distinguish land, water, and even forests. When their lead Cat broke through ice and settled into 6 feet of mud and water in a beaver pond they had been unable to see, the crew talked of quitting on the spot. Cables snapped as the men tried to winch out the Cat. Only after they blew the whole pond apart with dynamite could they retrieve the sunken bulldozer. Then the men refused to go any farther. That was when Miller put in the desperate call to Brown.

Tennessee Miller's determination to get a piece of the action on the North Slope was part of the frenzy created by Alaska's newest oil rush. The slope encompasses about 88,000 square miles of largely treeless tundra, and the betting was that this was where the next big fortunes would be made. Nearly every major company had joined the race.

The oil and gas potential of the region had been established shortly after the turn of the century. In his classic U.S. Geological Survey report published in 1919, E. de K. Leffingwell told of mapping the area from 1906 to 1914. He described and named Prudhoe Bay's unusual Sadlerochit formation, a sequence of sandstone and gravel deposited by an ancient river system about 220 million years ago. The heaving and thrusting of earth over time locked in deep formations of porous rocks, like a coral reef, which were capable of capturing migrating oil.

Oil explorers were also encouraged by the military's experience on the North Slope. When the U.S. Navy converted its ships from coal to oil power in the early 1920s, President Harding, heeding Leffingwell's words, looked to Alaska for the nation's future needs. He designated a 23-million-acre area at Umiat, just west of Prudhoe Bay, as a national petroleum reserve (NPR-4). The U.S. Geological Survey sent in teams to evaluate the potential, but drilling did not occur in the preserve until 1944, when World War II created a demand. After three dry wells, the Navy finally struck oil in Umiat. The reserves, estimated at 70 million barrels, were not enough to justify investing in a transport system, but they increased interest in Alaska's Arctic.

A Canadian heavy-drilling outfit was first to the slope when the ban was lifted in the early 1960s. Led by James C. Taylor, a rangy, fearless wildcatter, the Canadian team barged rigs and trucks 2,000 miles on the Mackenzie River to the Beaufort Sea and then into Alaska via the Colville River, which empties at Prudhoe Bay. They landed in the summer of 1963 at Pingo Beach, 40 miles upriver. They were promptly put to work by British Petroleum, among the earliest to buy leases in the North Slope sale. Taylor managed to move a drilling rig 60 miles by truck to the first BP site, earning the distinction of being the first to use wheeled vehicles on the arctic slope.

BP was certain it had a jump on all the other oil companies. However, it soon learned that drilling in the Arctic posed far more problems than drilling in the Iranian desert, where the company had excelled. Permafrost is often more than 1,400 feet deep in the Arctic. In the summer, the surface melts and the tundra becomes a soggy mess, flooding airstrips and making the ground impassable for wheeled vehicles. In the winter, when the ground is firm enough for drilling, the permafrost distorts seismic readings and can immobilize operations. One BP report complained that steel equipment fractured and normal lubrications solidified. The work was slow and frustrating.

But while BP proved it could drill deep into the Arctic, it could not locate arctic oil. The Canadian crew spudded wells at six different BP sites. All six turned out dry or with oil flows so weak they were useless. Nonetheless, experts kept insisting that the collective geological surveys, air photographs, and seismic findings indicated that the slope had potential oil-bearing structures comparable to Iran's giant oil fields. Such hype heightened the oil rush fever, and inspired Alaska to position itself for a bonanza.

Shortly after achieving statehood, Alaska began to take steps to reclaim leasing action from the federal government. In 1961, it started exercising the right to select a maximum of 103.5 million acres of federal land, which it had been granted under the provisions of statehood.

According to Charles Herbert, then commissioner of natural resources, the state's first choice was based on the advice of George Gryc, who had studied the arctic slope for the U.S. Geological Survey. "He thought the Arctic National Wildlife Refuge [the easternmost part of the slope] had the greatest oil potential, and I know the oil companies were pushing Governor Bill Egan to go for it," Herbert recalls. "But [the U.S. Department of] Defense wouldn't release it. However, they told us the government had no objections to us claiming Prudhoe Bay, which was next door."

The job of determining Alaska's North Slope holdings fell to land selection officer Tom Marshall. On the recommendation of advisers, he chose Prudhoe Bay. The site encompasses 4.2 million acres of the coastal belt that lies between the naval petroleum reserve at Umiat to the west and the Arctic National Wildlife Refuge to the east. Many Alaskans thought it was foolish to select land so forlorn that no one in his right mind would try to homestead there, but Marshall chose it with confidence. "This," he announced, "is where Alaska can make some money."

Next, the state adopted a system of competitive bidding. It announced that starting in 1965 it would hold a series of oil lease sales for its newly acquired lands. With that, the rush of exploration crews to the North Slope became fast and frantic. Anxiety ran high and rumors abounded about where on the slope the companies would stake their claims—and at what price. "Hell, they all came up here . . . California Standard, Phillips, Texaco, Chevron, Richfield, Tenneco, BP, and Pure. Yes, Shell was here, too," recalls veteran geologist Marvin S. Mangus, who mapped the slope for two decades. A Penn State graduate, Mangus worked for twelve seasons studying and mapping the area for the U.S. Geological Survey and then joined the Atlantic Refining Company in 1958 as a geologist in oil exploration. One group that Mangus remembers as conspicuously absent was Natives, despite the fact that the territory had been the ancestral home of Eskimos and Indians.

To the new explorers, who usually arrived at the onset of temperate weather, the North Slope tundra was breathtaking, a vast flat land abloom with wildflowers that seemed to go on forever. Wildlife roamed oblivious to man. Caribou, foxes, swans, and more than two hundred other species of birds claimed the spongy tundra, which was carpeted with millions of tiny plants of dazzling color. Only clouds of mosquitoes and leftover ice cakes on the pebbled shores of Prudhoe Bay reminded them of the harsh side of Alaska.

Mangus had long before reduced the character of the tundra to two impressions. "It is flatter than a bookkeeper's ass," he says. "And the caribou, the famous Porcupine herd, would come through around the twentieth of July—about twenty-five thousand of them. They'd mill like a damn barnyard until they left."

Mangus says he never doubted the slope's oil-bearing potential, even way back then—"Our USGS work had the structure and the trap. All we needed was the reservoir." Now, many of the USGS reports he had written provided a treasure map for his competitors in the race to find the reservoir. Ironically, little Richfield Oil, cash poor but savvy, showed more confidence in the USGS clues than Mangus's own company or the other industry giants.

In 1963, Richfield sent two young geologists, Gar Pessel and Gil Mull, to map the North Slope. Building on the data in the USGS reports, the two novice explorers were working along the Sagavanirktok River—Sag River, for short—at low water when they discovered oil-soaked sand exposed on an outcrop. "It smelled of oil," Mull recalls. "And the rocks seemed to have good reservoir potential."

On the next plane drop of supplies, Pessel sent a scribbled note back to Charles Selman, Richfield's division geophysicist in Anchorage. He described what he had seen and added, "If we can't find an oil field in this stuff, I give up." Selman took the note to Ben Ryan, the exploration superintendent, who enclosed it in a covering letter to Harry Jamison, his boss in Los Angeles, with a request to underwrite a seismic crew. Seismic testing—setting off explosives in deep-drilled holes and then analyzing the vibrations to map oil-bearing structures—is an expensive phase of exploration, but crucial to any drilling decision.

Jamison gave Selman a fast go-ahead. When he and his crew arrived at the slope, they found some unexpected assistance waiting. By some miracle, Tennessee Miller's Cat train had made it across the wilderness. Boyd Brown, who had replaced the crew chief, and his two other drivers, LeRoy Dennis and John Richart, arrived at the staging area to a rousing welcome, with bodies and machines reasonably intact.

To reach Sagwon, the men had slashed their way through densely thicketed ravines, bulldozed paths up and down steep, forested foothills, maneuvered across the uncertain ice of the mile-wide Yukon River, and cut a trail through a pass between 9,300-foot peaks on the Brooks Range. The 412-mile trip from where the first crew had stalled took twenty-two days, four of them spent bunkered down as gale winds blowing whiteouts made travel impossible. "The scariest part of the trip was crossing the frozen Yukon River," Brown says. "We put the first Cat in low gear and let it go across the river without a driver. The ice started cracking like someone firing rifle shots. But it made it, and the rest of the train followed, keeping some distance between us, of course."

Brown's Caterpillar-drawn sleds, complete with cookhouse and canvas-covered sleeping quarters, made the job easier as Selman's crew drilled and blasted across the tundra that winter. Today, of course, environmentalists would have thwarted such a ravage of interior Alaska before the bulldozers ever left Fairbanks. But this was

1964. Alaska's oil rush seemed to encounter no impediments then except the obstacles of the rugged state itself.

Oil companies closely guard the findings of their seismic explorations, but when the results of Selman's crew came back, Richfield decided to go against the prevailing wisdom. The possibility of large oil-bearing deposits seemed so great that the company sensed it was on the verge of a find much bigger than its limited resources could develop. It couldn't swing it without a well-heeled partner. That meant sharing its secret with a competitor.

Harry Jamison, Richfield's hard-driving exploration chief, flew to Alaska and picked up a sample from the oily outcrop located by Mull and Pessel. He took it to Los Angeles and carried it under his arm into the office of J. R. Jackson, manager of West Coast exploration for Humble Oil. Jamison laid Richfield's findings before Jackson and invited Humble to become an exploration partner in Prudhoe Bay. "We had the competitive jump, and Humble had the deep pockets," Jamison says. "It was the kind of arrangement that could work out for both of us."

Humble had already given up its own exploration in Alaska after drilling a series of dry wells. But the glow on Jamison's face was convincing. Consultations with Humble's board in Houston followed, and a deal was struck. Humble agreed to share half of the expenses. That meant putting $1.5 million into more exploration, paying half the cost of lease purchases, and assuming half of the expenses of wildcatting—a total investment in Prudhoe of $40 million for both companies.

When the state held lease sales for Prudhoe Bay tracts on July 14, 1965, Richfield surprised the industry—and just about everyone else. It bid an unheard-of $93.78 an acre for tracts on the crest of the Sadlerochit formation at Prudhoe Bay, outbidding everyone for about two thirds of the structure and pouring more than $1.3 million into Alaska's treasury.

Meanwhile, British Petroleum had seismic data on Prudhoe Bay that looked promising, too. But with its development funds sapped by a series of drilling failures on the slope, it did not have the money to compete with Richfield. Nevertheless, BP managed to become a major player in Alaska by gambling on the flanks of the structure, which it acquired at the relatively low bid of $47.60 an acre. Ironically, Atlantic Refining, Mangus's company, was outbid on every parcel and did not nail down a piece of Prudhoe Bay.

Strengthened by the infusion of Humble money, Richfield shipped a complete set of drilling equipment, some three thousand tons of rigs, camps, and pipe, to Fairbanks. Then it arranged for a Hercules C-130 four-engine air-freight plane to fly the equipment to a drilling location on the slope.

However, before Richfield could spud its first well at Prudhoe Bay, a problem simmering in Washington clouded the company's bright prospects. The antitrust division of the Justice Department had started to sue Richfield three years earlier. Now it was turning up the heat. It sought to break up Richfield's overlapping ownership with Cities Service and Sinclair Oil, an arrangement in effect since the mid-1930s.

Cities Service and Sinclair had provided the financial support to bail out Richfield when the oil company went into receivership during the Depression. Each received about 18 percent of the stock of the reorganized Richfield in 1937. The plan had not only been approved by the U.S. Circuit Court of Appeals, but had also passed the scrutiny of the Sabath Committee, the House Select Committee to Investigate Bondholders' Reorganizations. So why, after twenty-five years, was the Justice Department suddenly intent on forcing Richfield to divest?

On March 1, 1962, W. Alton Jones, president of Cities Service (no relation to Richfield president Charles Jones), had been killed in an American Airlines crash over New York. Investigators found that he was carrying $55,690 in cash, most of it in $1,000 bills packed in a black leather bag. The seventy-year-old millionaire was on his way to California for a fishing trip to Mexico with his friend, former president Dwight Eisenhower, according to company spokesmen. Jones was also a well-known supporter of Republican causes, and suspicions quickly grew that he had been on his way to California to deliver a secret contribution to Richard Nixon, who had been defeated by John F. Kennedy in the 1960 election.

The money in Jones's possession got the attention of Robert Kennedy, the president's brother and at that time U.S. attorney general. He was among those who believed that Jones had been on a mission to help finance a Nixon comeback. Bobby Kennedy's inclination to use his office to go after industrial fat cats—particularly Republicans—had been well demonstrated. Early in 1962, he shook up the steel industry with an attempt to convene a grand jury to investigate price-fixing by the leading steel company executive officers—good Republicans all—after President Kennedy castigated them for raising prices. Now he seized the chance to make big Republicans in the oil industry uncomfortable. The Justice Department started the antitrust suit against Richfield in October 1962, about seven months after W. Alton Jones was killed. Charles Jones, Richfield's chief executive, decided to fight, but by 1965 he faced a catch-22. Contesting the case to the end would run up tremendous legal fees that would weaken the company and surely imperil its ability to proceed in Prudhoe Bay. Yet the only way to buy out Cities Service and Sinclair would be to sell or merge, which meant he would lose his company.

Jones reluctantly agreed to hunt for a merger partner so he could buy out the two minority partners. "Then a funny thing happened," Jones later wrote in his autobiography. "One of our Washington lawyers wrote to tell me that a Mr. Cladouhos, an antitrust division lawyer assigned to the case, had suggested that Richfield settle the suit by merging with Atlantic Refining Co." Why, Jones wondered, was the Justice Department recommending merger, and why with Atlantic Refining? Jones never got his answer, but he was astute enough to sense that it would be prudent to follow the advice.

Atlantic Refining Company, then the nation's thirteenth-largest oil enterprise, was based in Philadelphia and had assets of $900 million. Its chairman, Robert O. Anderson, had been described in *Business Week* as a "loner with a Midas touch" who possessed "an incredible sense of timing." He also was America's largest individual landowner, controlling about 1.02 million acres of cattle ranges in New Mexico, Texas, and Colorado.

The merger between Atlantic Refining and Richfield Oil occurred on September 16, 1965, much to the distress of many proud Richfield employees. They felt they were on the verge of beating the industry to one of the world's great oil finds. Now, Atlantic Refining, which had been outsmarted at just about every stage of the race to the North Slope, stood to become a major beneficiary of Richfield's daring and talent.

On February 27, 1966, the new company, now known as Atlantic Richfield, drilled its first well at Prudhoe Bay. Dubbed "Susie," the site was in the foothills of the Brooks Range, well south of the arctic coast on the edge of the Sadlerochit formation. Despite all the high expectations, it turned out dry.

"It was a blow," Charles Selman, the seismic crew leader, recalls. It came on top of the experience of BP and Sinclair, which had drilled six wells—all dry—within the last eighteen months. The luster of the North Slope was now fading fast. Skeptics at Atlantic Richfield urged that no more money be spent on this remote region of Alaska, pointing out that all the other majors had given up. "After Susie failed, we were the only act in town," Harry Jamison concedes. "No one else was drilling."

But Jamison was unwavering in his belief. He put his career with the merged company on the line, insisting that they drill just one more hole. His confidence impressed Thornton F. Bradshaw, the Atlantic executive who had become president of the reorganized company. Bradshaw persuaded the board that one more well would not require a new infusion of capital. The company would have to pay for dismantling the equipment and flying it out to Fairbanks in any event. Why not drill a second well while it still had an assembled drilling rig sitting on the slope?

A mixed team of Richfield and Atlantic oil searchers met in Los Angeles to pick a drilling site. Jamison, Selman, district geologist Don Jessup, and district exploration manager John Sweet studied the strata on a map drawn by Rudy Berlin, a geophysicist, and reached a consensus. "After drilling Susie, we were sure that oil had migrated farther north," Jamison says. "We decided that it had left the marine soils of the cretaceous field in the south and was trapped in the Paleozoic, an older and different kind of rock formation that characterizes the north of Prudhoe Bay. That's where we pinned the new location."

Tennessee Miller's Cats moved the drilling rig 65 miles to the north, almost to the shore of the Arctic Ocean, and drilling on the new well began on April 22, 1967. Marvin Mangus, Atlantic's veteran slope geologist, and Gil Mull, Richfield's young geologist, were among those who took turns "sitting on the well," monitoring the samples of soil, rocks, and moisture as drilling progressed.

The first big break came on Mull's watch. "It was the day after Christmas, 1967," Mull says. "We opened a hole to let the fluid flow to the surface. Out came a tremendous burst of gas. It ignited at the end of a two-inch flow pipe and it blew a fifty-foot flare into the teeth of a thirty-mile-per-hour wind. A gas flow doesn't necessarily mean you have liquid, but that kind of pressure is characteristic of a major reservoir." Two months of drilling later, samples at 8,820 feet had all the signs of a major oil find. Harry Jamison, now bearing the title of Atlantic Richfield's general manager for Alaska, was confident enough to call in the press on February 16, 1968. "It looks extremely good," he said. "We have a major discovery."

Alaskan newspapers splashed the big oil strike in sixty-point headlines. The residents of Fairbanks—the closest city, 390 miles to the south—turned out to celebrate in the streets. Governor Walter J. Hickel, then serving his first term in the capital, hailed the "great news" and predicted big enterprises to come.

At first, Wall Street was reasonably calm. Stock analysts cautioned investors that the find might be a fluke. A confirmation well was drilled on a site 7 miles away on the Sag River, the area where young Richfield geologists Mull and Pessel first spotted oily sand. Mull was the "sitter" on this well, too, and once again spectacular things happened on his watch. "It shot out a full column of oil," Mull says. "This meant that the entire huge formation underneath had oil." Prudhoe Bay was no fluke.

No one shared in the joy of discovery more than British Petroleum. P. E. Kent, chief BP geologist, concedes that the company had been ready to cut its losses and leave Alaska. British Petroleum had finished dismantling its equipment and was seriously considering an offer from Atlantic Richfield to take over BP's Prudhoe acreage when news of the first discovery came in. Then, as Kent put it, "our technical assessment did not encourage easy capitulation to our main competitor."

By holding out a few months longer, BP saw the second Atlantic Richfield well come in, but it took a year more to realize the full magnitude of the decision. And the realization was a stunning one: When BP finally struck oil in March 1969, it found that its leases on the flanks of the Prudhoe Bay structure would contain the greater share of oil deposits in what became North America's largest oil field. Five smaller companies also discovered their first North Slope oil soon after. Mobil came in with a find that would enable it to become an 8.5 percent partner in building a pipeline system; Phillips and Union Oil each qualified for 3.25 percent, Amerada Hess for 3 percent, and Home Oil of Canada for 2 percent. The combined companies estimated that this 350-square-mile field at Prudhoe contained about 22 billion barrels of original oil and 30 trillion feet of natural gas in an overlying gas cap.

This time the news from Alaska created tremors on Wall Street. Disregarding all cautions from analysts, speculators rushed to buy stock in the discovery company. Atlantic Richfield stock shot from $90 a share to $162 during the five months from discovery to confirmation. Action became furious and rumors ran thick on Wall Street. *New York Times* writer Robert S. Wright tried to call for calm in his "Market Place" column on July 10, 1968: "Fantastic reserve figures [in the Alaska discovery] are making the rounds of Wall Street. The Umiat Chamber of Commerce, if there is one, could not be reached for comment . . . but even if the North Slope discovery turned out to be a major source of oil, many problems and many years would be required to get the petroleum to market."

Wright was more than correct. The problems to come would be greater than anyone envisioned.

One obstacle that was already becoming apparent was the impact of oil development on Alaska's environment. Soon after the discovery at Prudhoe, in the name of helping to develop the state's new leading industry, Governor Hickel slashed a highway to the North Slope. The ill-planned road followed much of the route bulldozed by Boyd Brown and his Cat train. But Boyd and his crew at least had cut their road in winter conditions, so they had not damaged the permafrost. Nature would have healed the surface wounds in time. Hickel, however, tried to make the highway an all-season road by bulldozing into the permafrost. That made the damage permanent. The sun melted the surface in summer, turning the road into a canal, and without a bulldozer to buck the drifts, it remains impassable in winter. The road became derisively known as "Hickel Highway" and is seldom traveled today.

"We honestly believe Hickel built that road just so a contractor pal of his in Alaska could get his equipment up to the slope ahead of the others and get a jump on the business," says Celia Hunter, a pioneer conservationist who helped form the Alaska Conservation Society, the first statewide environmental watchdog group. "There were no permits taken out to build the road, no public hearings and no chance to be heard. Hickel just built it.

"So much environmental damage was done without public knowledge," she says bitterly. "We couldn't catch up to it until after the fact. Then all we could do was try to spread the word in our newsletter."

The problems would become even more complex. The threat to Alaska's wilderness was growing larger than arrogant governors and reckless Caterpillar skinners. The biggest names in the oil industry were readying to skim the riches of a land they didn't even own.

CHAPTER 5
The Settlement

ONE EARLY spring day in 1962, three Athabaskan Indian trappers set out from Tyonek, their remote, poor Native village on the west side of the Cook Inlet, to search for signs of lynx and fox in the thick forest that lines the rugged western shore. As they climbed a knoll of black spruce and birch about 6 miles from the village, they came upon a sight that stopped them in their tracks.

There, in the midst of a newly cleared plot of land on their ancestral hunting grounds, stood an oil drilling rig with several white people moving about it. "What the hell goes on?" Bill Standifer whispered to his companions, George and Alec Constantine. The three Natives dropped their traps and ran back to alert the village that white men were stealing their land.

Emil McCord, a muscular, bushy-haired officer of the Tyonek Village Council, quickly rounded up a band of villagers, vowing to drive the intruders off the reservation. They marched to the site and surrounded the rig. Emil climbed a ladder to confront the drilling crew. "What are you doing here? This is our land!"

"It's none of your fuckin' business," a driller shouted back. "Get off the equipment or we'll throw your ass off."

Emil's inclination was to storm the drillers and bash the one spewing obscenities. However, cooler heads calmed him down. "We decided to go back to the village and radio for Stanley," he says.

Stanley McCutcheon was a white Anchorage attorney with a social conscience, a liberal activist who understood service to country well before President John F. Kennedy's call made it politically fashionable. A bush pilot as well as a lawyer, McCutcheon would voluntarily fly a relief plane from Anchorage during the harsh winters. When word went out that the people of Tyonek were starving, Anchorage grocery stores would set aside barrels to help the Indians; people deposited canned goods and packages of food, which McCutcheon delivered.

Despite the repeated threat of starvation, the 260 villagers preferred poverty to giving up their traditional subsistence life-style. Their ancestors had hunted, trapped, and fished in the area for centuries, handing knowledge down to each generation about the patterns of the creatures that inhabited the forest and rivers and the ways to outwit them. In 1915, President Woodrow Wilson sealed their right to the land by signing an order setting aside 24,000 acres as the Moquawkie Reservation, one of the few ever created in Alaska.

The reservation had no roads or railroads leading to it. For years, the Tyoneks lived there in peaceful isolation, taking the moose, salmon, and other resources the land gave up and enduring the hardships when it did not yield enough. With more people exploiting the shrinking frontier, however, subsistence living became tougher over the years. Moose and furbearing animals grew scarcer, and in 1952 the government sharply cut back the season for commercial salmon fishing, the town's main source of cash. By the 1960s, Tyonek had turned into the Appalachia of Alaska.

Francis M. Stevens, a BIA child welfare specialist assigned to Tyonek in 1960, found the people living in hovels, sometimes as many as ten crammed into drafty, one-room tar-paper huts. The only water supply was a communal spring. They also had to carry out their body wastes, since the village's only flush toilet was in the school, a run-down building maintained and manned by the BIA. There was only one store in town, run by the BIA and stocked once a year by its supply boat, the *North Star*. Tyonek became a company town, and the company was the dreaded BIA, the subgovernment within the federal government that operated the welfare programs for Natives in Alaska.

Stevens was appalled by the federal bureau's insensitivity to the plight of the troubled Natives. It often removed children from their parents' custody without court process, usually on the mere word of a public health nurse or a BIA teacher. The children would be placed in group care facilities run by "Bible-thumping fanatics from outside," as Stevens described them. They collected $250 a month per capita from the government for housing as many as forty children at a time. "These were pretty decent white people, but they all had some kind of hard religious bent. They came to Alaska to do good, but they did well," Stevens observed after seeing the government payments.

Stevens, who held a graduate degree in social work from the University of Minnesota, challenged the BIA's bureaucratic smugness. At one point, following a drunken bash in the village, Stevens was directed to tell the villagers they would get no more welfare until they stopped drinking. "I refused to do it," he said. "I was willing to go along and say, 'No more cash welfare,' but cutting off food and bringing starving families to their knees was not my way of dealing with the problem."

Now, the appearance of the mysterious oil rig on Tyonek land was about to force a showdown that would challenge the BIA's long hold on the reservation.

When Stanley McCutcheon heard about the rig, he saw an opportunity to help his friends obtain a measure of retribution. Ever since the first oil field had been discovered at Swanson River on the east side of the Cook Inlet in 1957, oil companies had pushed their explorations westward. Productive oil and gas wells were found in the inlet and new ones were discovered ever closer to the western shore, where the Tyoneks lived. The reservation was the next obvious step in exploration.

McCutcheon went to federal district court without even discussing a fee. He expected none, and, in fact, the village had only $14 in its treasury at the time, according to Emil McCord.

McCutcheon succeeded in exposing the BIA's extraordinary highhandedness and insensitivity. In 1962, without notifying the villagers, the BIA had leased a section of the Tyonek reservation to the

Pan American Oil Company (since merged with Amoco). It placed the money in a government fund, also without telling the village. A swift court injunction stopped the drilling. Then McCutcheon sued the Department of the Interior, parent agency of the BIA, to prevent it from executing oil and gas leases on the Moquawkie Reservation without consent of the Tyonek Village Council.

The victory was stunning. The BIA was barred from making further deals without the village's permission, and it was forced to loosen its grip on any village money held in trust. Tyonek had finally regained a measure of control over its own future. With an embarrassed BIA doing the paperwork, the village proceeded to renegotiate all oil exploration leases on its reservation.

Because of Tyonek's proximity to the new oil finds in the Cook Inlet, major oil firms bid eagerly and heavily. The new leases generated about $12.5 million of the $14 million that would eventually swell the village treasury to undreamed-of wealth. This was the same land that the BIA had been willing to lease for $1 million.

The business acumen of the Tyonek Natives exposed the BIA's incompetence. Once the administrator of education, health care, and welfare for all Alaska Natives, the BIA had been reluctant to dismantle its bureaucratic empire, even though many of its services were to be transferred to Alaska under the provisions of statehood. The Tyonek incident provided new reason to hasten the transition. "All of us on the council went to Anchorage and rented rooms at an old hotel on Fourth Avenue," McCord recalls. "Then we all got drunk toasting our victory over the BIA—the Bastards Inflicted on Alaska."

The Tyoneks used their newly gained wealth to replace their hovels with modern houses, to erect a school as up-to-date as any in Alaska, and to ship tons of rice to poor villages in the north. Sadly, the boom was short-lived and ended as soon as the money ran out. Today Tyonek is struggling to keep its culture intact, but it takes great pride in its victory over the BIA and in its status as an exclusively Indian village.

Willie Hensley saw even greater rewards in the Tyonek victory. Here, he sensed, was a foot in the door to a further restoration of Native rights.

Up to now, the issue of Natives' rights to their land had been largely dismissed by most Alaskans as a local battle of vested interests: homesteaders and miners staking claims in the wilderness versus Natives claiming encroachment on their hunting grounds. Over the nearly one hundred years that Alaska had been an American territory, much lip service and vague language had been given to the rights of aboriginals, but it was not until 1959 that Natives received the first tangible indication that they had a right to recover for the loss of property or rights in property.

That breakthrough occurred in the United States Court of Claims, which ruled in favor of the Tlingit and Haida Indian groups in Alaska's southeast. It awarded them $7.5 million for the 20 million acres that the government had appropriated to create the Tongass National Forest and Glacier Bay National Monument in the early 1900s. But the Tlingit-Haida claim had lingered in the courts for more than twenty years. The Tyonek victory had taken less than three years, and it had broken important new ground.

McCutcheon had won by demonstrating that the Interior Department's administrative policies were oppressive. He showed that the Tyonek Indians had been treated as second-class citizens when such treatment was being repudiated across the United States in the civil rights revolution. The case had brought national attention to Alaska Native rights. With the country's mood shifting to their side, Willie Hensley felt it was time for Natives to fight the white man with his own techniques.

Willie enrolled in graduate school at the University of Alaska Fairbanks in 1966. He spent much of the year researching a thesis he titled "What Rights to Land Have the Alaska Natives?" The premise was that no treaty or act of Congress had ever extinguished Natives' title to their land. Yet Alaska had gone about selecting the 103 million acres granted under the Statehood Act of 1958 without considering Native rights to those lands.

Unless help came soon, Willie feared, Native lands in Alaska would go the way of Indian lands in the lower forty-eight. It is a sordid fact of history that the government appropriated or sold more than 90 million acres of Indian property between 1887, when treaty-making stopped, and 1934, when the Roosevelt administration managed to halt further land grabs. "We felt that was our land and most Native people felt it was their land," Willie says. "But we had no idea of what transpired in the past . . . what legal, political activities had taken place. We didn't know anything about protecting our ownership."

In the spring of 1966, while baby-sitting for a friend from college, Willie saw a story in the *Fairbanks News-Miner* that inspired him to raise the issue publicly. Alaska's Senator Ernest Gruening, alarmed that some Natives were protesting state land selections around their villages, was quoted as saying the government probably ought to pay off Natives so the state could get on with its development. "Here, finally, was a U.S. senator agreeing that we deserved some compensation," Willie thought to himself. "So that means he must agree that we have some rights to the land. And if we have a right, then we have a right to say what kind of settlement we want."

The next day Willie sent a letter to Senator Gruening and mailed copies to all major Alaska newspapers. It read:

> Dear Senator Gruening,
>
> You cannot pin the responsibility for the present chaotic state of affairs of Alaskan lands on the Natives. There has been ample time since the Treaty of Cession in 1867 to interpret its provisions. Now that we are finally taking limited action on an issue which has been allowed to ride, I pray that the claims will be allowed to be heard and a just and equitable settlement be made in real property or in cash, depending on comprehensive hearings with full participation of Alaskan Natives.

The newspapers reprinted the letter and Senator Gruening promptly invited Willie in for a talk. Willie was scared, but he welcomed the chance to press his case upon one of the most prestigious and powerful figures in Alaska.

"We met at the Nordale Hotel in Anchorage, in the room where he lived, just the two of us," Willie says. "I told him of my research, that our title to the land had never been extinguished, either in the purchase treaty with Russia or in the statehood agreement. And therefore we had a right to say what kind of settlement we wanted. He was very nice about it. He said he was a friend of the Natives and reviewed his record as territorial governor in which he fought racism in Alaska and helped get rid of those signs in theaters and restaurants that had segregated Natives."

Willie said he didn't question Gruening's position on racism, but he felt a chill when he got to specifics on Native land claims. "I feared he wanted to do the traditional thing—go to the court of claims or the Indian Claims Commission, pay the Natives fifty cents an acre, and then move on. He had no feeling that we wanted to retain land." Gruening made no promises, except to say Natives deserved something. Willie left the meeting feeling that Alaska's more liberal senator was more a colonialist than a champion of Native rights.

Willie graduated with his master's that spring, with high praise for his thesis from one of the university's most respected teachers, Judge Jay Rabinowitz. Then he borrowed the money to return home to Kotzebue. There, still virtually penniless, he declared his candidacy for the Alaska legislature. At twenty-five, Willie had grown into distinguished adulthood. He was trim and neat, his manner both engaging and refined. And no one could speak more authoritatively—or persuasively—on the problems facing Alaskan Natives.

Borrowing $10 for paper and stamps, Willie wrote to all village leaders in the Kotzebue–Yukon Delta area, his legislative district, which has one of the greatest densities of Natives in Alaska. He invited them to a meeting, telling them that land their people had lived on for centuries might soon pass to state ownership and then be offered to public sale or lease.

"A lot of people didn't read too well," Willie says. "One guy from Noatak came seventy-five miles by boat to ask what the letter was all about." However, a strong local showing from Kotzebue enabled Willie to organize a Native group, which its members named the Northwest Alaska Native Association. Willie was chosen its first executive director. Now he had a forum and base. At the group's third meeting, Willie asked, "What should we do about our land?" "Claim it!" a chorus shouted.

Very little land in Alaska had actually been surveyed, despite the leasing frenzy. So Willie took a big, black pencil and drew lines on a map, circling what he felt was the drainage area of the Kotzebue region, and submitted the claim to the Bureau of Land Management and the Alaska land office.

There had been another Native land claim earlier that year, by the Arctic Slope Native Association. The group had been organized also on the spur of the moment by Charles Edwardsen, an unkempt, hard-drinking young Eskimo from Barrow. He claimed title to 96 million acres, which consisted of practically all of Alaska north of the Brooks Range and covered all the potential oil fields the oil companies were exploring. A product of Mount Edgecumbe, the BIA high school in Sitka, Edwardsen had the smarts to go to college, but had dropped out. Copying the style of civil rights demonstrators of the era, he became an Eskimo radical, decrying "white trespassers" and suggesting that violent means might be necessary to save the land. Edwardsen wrote to Willie urging that the Kotzebue Natives team up with their fellow Inupiats in the north. But Willie was not ready to endorse Edwardsen's brand of activism. He felt he could win only if he played by white man's rules.

Willie made land claims his campaign issue in Kotzebue and then carried his crusade to Anchorage, hoping to find support for making it a statewide concern. There he met Emil Notti, one of the few other Natives in Alaska with a college education. Notti, short, dark-haired, and half Athabaskan, has scars from his left cheek to his neck, remnants of an attack by a dogsled team when he was growing up along the Yukon River in the interior town of Ruby. Four years in the Navy provided him with his escape into the outside world, and the GI Bill enabled him to earn an engineering degree from Northrop University in California. Notti had sensed social injustices to his people ever since a BIA school taught him how to read. "You would see pictures of famous people coming to Alaska, such as Roy Rogers, and they would have a picture taken with their trophy bear or moose," Notti says. "Inevitably, the Native person in the picture would be identified only as 'the Indian guide.' No name, even though he probably located the animal or even shot it for him. The Indians didn't have names, so you assumed they were not a person."

When he met Willie Hensley, Notti already was organizing a Native association in the Cook Inlet region, covering principally southwestern Alaska. Both men agreed that only a statewide organization of Natives could stop the seizure of their lands, but neither of their organizations had the staff or the money to assemble village leaders from across Alaska. Most Natives lived in villages with no access roads—although Alaska is more than twice the size of Texas it had only 6,500 miles of roadway—and to fly them in would cost too much money. Few villages could underwrite trips of such distances for their leaders.

However, both men had heard of the deeds of Tyonek. "Emil Notti and Willie Hensley came down and said they needed money," Emil McCord recalls. "They said Natives had to start doing things on their own or we were going to lose our land. We loaned them a hundred thousand dollars with a handshake." The Tyonek Village Council also agreed to pay the travel and room expenses for the delegates and arrange a meeting place in Anchorage as well. With a barrage of publicity and editorial support from Howard Rock, the revered editor of the *Tundra Times,* the weekly publication for Natives, the prospect of Alaska's first Native summit meeting became real.

However, the prospect of a consensus was something else. Alaska's Natives were isolated, splintered, and poorly educated. Because of the vastness of the state, communication among groups was possible only where borders touched. The profusion of languages and dialects alone threatened to doom any statewide conference to a cacophony of confusion. Nevertheless, on October 6, 1966, three hundred Native delegates from all parts of Alaska packed a deserted storeroom above Miller's Furs on Fourth Avenue in Anchorage, and the Alaska Federation of Natives (AFN) was born.

The anticipated chaos did not occur. The determination to save their lands overcame the groups' differences. The conference coalesced into a mission in which every Native group would have a role. Flore Lekanof, an Aleut from the Pribilofs, became permanent chairman. Robert Peratrovich, president of the Tlingit-Haidas in the southeast, headed the education committee. Richard Frank, leader of the Tanana Chiefs Conference from the Athabaskan interior, was named a director. Even the radical Charles Edwardsen made the team as chairman of Native employment.

The conference was held virtually on the eve of statewide elections, and political candidates swarmed among the delegates seeking votes. It was an eye-opener for many of the Natives, who were getting a speedy lesson in election clout. Alaska had fifty-five thousand Native votes to court.

Emil Notti, the first AFN president, quickly created a land claims committee and appointed Willie Hensley to head it. Then the committee drafted a position statement that was released to the press and sent to Alaska's congressional delegation. The statement contended that all Alaska Natives were being victimized and they wanted reparations.

Most whites in Alaska reacted indignantly to the news that Natives were claiming land title and money as compensation for acreage being selected by the state. Although the position paper stated that AFN only wanted input in the settlement, reports spread that Native leaders were demanding 60 million acres and a billion dollars. Governor Hickel added to the public furor, declaring in a statewide radio address, "Just because somebody's grandfather chased a moose across the land doesn't mean he owns it."

Months went by and still no member of the congressional delegation reacted to the AFN position paper. In April 1967, Notti sent telegrams to all three members of the Alaska congressional delegation. Senator E. L. Bartlett responded immediately. Senator Gruening responded two weeks later, and Representative Howard Pollock's response came shortly afterward. They agreed to place the Native claims issue before Congress in the form of Senate bill 2020. It was the first of eight bills that would be introduced before Congress finished its own soul-searching on the lands issue.

Meanwhile, Willie Hensley had been handily elected to the state legislature, defeating four strong candidates in the primary and a respected Republican in the fall election. He promptly used his new position to fight the state's arbitrary land selection. Stanley McCutcheon, his legal expenses paid by Tyonek, traveled to villages from Nome in the Arctic to Kodiak in the southwest to press the same cause, joined by his law partner, Cliff Groh, Sr. They helped Native organizations become legal entities so they could file land claims and block state selection of their regions.

By 1967, Native organizations had claimed more land than there was in the state of Alaska, completely stymieing the state's process. In response to this chaos, Secretary of Interior Stewart L. Udall declared a land freeze for all Alaska. He refused to accept further selections by the state until the Native claims issue was resolved.

The freeze halted all development. Cries of "greedy Natives" were interspersed with various stories of individual hardships caused by "federal tyranny." Many Alaskans still insisted that Natives had no right to compensation whatsoever. The powerful *Anchorage Times* editorialized that land and money demands being made by the Natives would cripple development for all Alaskans. The devastating economic effect of Udall's edict drove Governor Hickel, who was counting on land sales to balance his budget, to desperate moves. He tried to lift the freeze by suing Udall in federal district court. Then he attempted to get a quick fix in Congress. Nothing worked.

Udall's genuine sensitivity to Native claims and the national mood charged by the Johnson antipoverty pledges and the civil rights demonstrations of the 1960s were big pluses for the Native cause. But it is doubtful whether the Native land claims would have survived the hostility in the state during this era had they not coincided with the epochal discovery of oil at Prudhoe Bay in 1968.

Suddenly, the oil industry became an overpowering—if unlikely—booster of Native rights. Realizing it could not tap the massive oil field on the North Slope without a clear land title, the industry applied its considerable muscle to Congress. Newfound concern for Native rights in Alaska suddenly started emanating from Governor Hickel, from Congress, and, starting in 1969, from the inner sanctum of the Nixon administration. John Ehrlichman, who as a Nixon campaign strategist had discovered the advantages of helping the oil industry, was particularly emphatic. Congressional library shelves bulge with transcripts reflecting the millions of words that went into the deliberations on Alaska Native land claims bills, including a presidential commission field study, the testimony of scores of witnesses called before the Senate Committee on Interior and Insular Affairs, and letters and reports from interested parties ranging from environmentalists to sportsmen's councils. Even with the high-level response to the oil industry's urgent call, the Native land claims settlement was a tortuous process.

Walter Hickel developed a change of heart the hard way. As governor, he could afford to oppose the land freeze and vacillate on the land claims issue. However, when President Nixon nominated Hickel as his secretary of the interior after the 1968 election, Hickel needed Native support to counter the outrage among environmentalists who protested that his appointment would be like installing a fox to watch the henhouse.

Emil Notti, then president of the Alaska Federation of Natives, recalls being asked to visit Hickel's home late in December 1968. He was driven there by Cliff Groh, the AFN's attorney. They found a troubled Governor Hickel with Larry Fanning, editor of the *Anchorage Daily News.*. Fanning was holding an advance copy of a column by Drew Pearson, Washington's leading muckraker and then the nation's most widely syndicated columnist. Pearson had written that Hickel did not deserve to be secretary of the interior because of his shameful treatment of Eskimos in Bethel, a western Alaska Native community on the Kuskokwim River. Clearly, Hickel needed to blunt the attack.

He asked Notti if he had contact with Indian groups. Notti replied that he had. "Do you think you could get an endorsement from them?" Hickel asked. Notti said he would like to be able to help, but he reminded Hickel of their running battle with him over the land freeze. What about an endorsement from the AFN? Hickel wanted to know. It would be the same with that group, Notti responded.

Notti says Hickel then promised to extend Udall's land freeze until Native claims were resolved. Notti wanted the promise in writing. Wasn't his word good enough? Hickel countered. Notti said that was not good enough, and left.

Groh was furious when they got to the car. "You called the governor a liar!" he exploded. Notti reminded Groh of Hickel's recent statement at a Seattle airport press conference: "What Udall could do with a stroke of a pen, I can undo with a stroke of a pen when I become secretary."

A few days after the meeting at Hickel's house, Notti called a special meeting of the AFN board to act on Hickel's plea for an endorsement. He recalls that a three-way phone call was set up with Hickel on one end, Cliff Groh on another, and a speakerphone in the board meeting room. The conversation ended without a commitment from the board. Apparently, Hickel thought the board had signed off, because suddenly the governor's angry voice came blaring over the speakerphone: "Cliff, you made a promise. When the hell are you going to deliver them?"

"That just jolted us," Notti says. "I had thought about going Outside for counsel earlier because of the way Congress jerked our local lawyers around. Hearing that exchange between our lawyer and Hickel convinced me we had to reach Outside."

Notti said the AFN sent him to the Hickel confirmation hearings before the Senate Committee on Interior and Insular Affairs with the authority to endorse, withhold endorsement, or oppose, depending on what Hickel said about maintaining the land freeze until Native claims were settled. Hickel dodged questions on the freeze for the first two days of the hearings. At the end of the second day, January 18, 1969, Senator Henry Jackson, chairman of the committee, put the question to Hickel directly:

> Governor, during yesterday morning's session, I asked you a number of questions concerning your intentions on dealing with the land freeze in the state of Alaska. Now as I understand your answers . . . we had arrived at a clear understanding that no action would be taken by you as secretary of interior to lift the land freeze or otherwise change the status quo. . . . I have been informed that while I was out of the meeting yesterday you responded to the questions of Senator [George] McGovern and other senators on the same subject. The concern has been expressed that in your answers you may have modified somewhat the understanding which I thought we had attained. I would appreciate it if you would clarify this matter for me once again.

Hickel faced his worst moment in the hearings. As governor only a few months before, he had sued the Interior Department to lift the freeze. He had hoped to parry questions and leave himself an escape. Now he knew that his confirmation rode on how he satisfied Jackson. Yes, he said, Jackson's first understanding was correct and he would abide by it. That statement under oath was also enough to wrest the endorsement of the AFN.

Notti and Hensley estimate they and other Native leaders made about 120 trips to Washington to protect Native interests as the Alaska delegation negotiated and Congress debated land claims bills. More tiring than the traveling and negotiating, though, were the growing internal disagreements that threatened to shatter the fragile AFN. The disputes were fanned by a battery of lawyers who, sensing a gigantic legal payout, maneuvered to get a piece of the action. Notti says that at one point fourteen attorneys, each representing a separate region, were advising their clients what to demand to satisfy the needs of their regions in any land claims settlement act. Each attorney, of course, was expecting a slice of the settlement.

Many of the provincial demands were unreasonable, but pertinent issues also emerged. Emotional debates developed over protection of hunting and fishing rights, over how land shares would be apportioned, and over the diversion of funds from housing programs to pay for the mounting expenses of the claims fight. However, at the root of the friction was the struggle over which local attorneys got to represent the AFN in the congressional hearings. "These guys were working for ten percent of the land and the money in any settlement," Notti said. "They stood to make a lot of money."

Robert Goldberg, a young attorney who came to Alaska in 1967 and landed a job with an Anchorage law firm, told his father, former Supreme Court justice Arthur J. Goldberg, about the Native concerns and about the many attorneys trying to board the Native claims gravy train. He asked his father to help.

Goldberg had visited Alaska and had grown fond of the state. After confirming his son's reports, Goldberg offered to represent the Native claims without pay. Later he was joined by former attorney general Ramsey Clark, also working on a pro bono basis. Together they saved the AFN from possible disintegration and provided immense credibility for the Native case. "Up until then, we were a bunch of wild Natives trying to claim Alaska," Willie Hensley says. "Their defense of our cause made an impression on everyone—the Senate committee, the media, and the public. Keep in mind that just a short time ago, Ramsey Clark as attorney general was on the other side of the fence, fighting against Indian claims. His willingness to work for us made quite a statement."

On April 29, 1969, Goldberg made a masterful opening statement at the hearings of Senator Jackson's Senate Committee on Interior and Insular Affairs: "The Alaska Natives came to their nation—and it was their nation at the time—thousands of years ago. By sheer determination and ingenuity, they have endured, indeed they have conquered, an arctic semicontinent without the benefit of modern technology, without electricity or planes, or matches or modern heating equipment. They survived under conditions which might have destroyed any other group in the world."

Then Goldberg considered the plight of Alaska Natives in relation to the national War on Poverty, a comparison that struck a nerve in Congress: The average Alaska Native died at the age of 34.3 years, and thus had half the life expectancy of other Americans; death for Natives from influenza and pneumonia occurred at twenty times the rate for Alaskan whites; fifty-two out of every thousand Native Alaskans died before reaching their first birthday, and the Native infant mortality rate was twelve times that of white Alaskans; more than half of the Native work force was jobless most of the year. And out of some seventy-five hundred residential dwellings in Native villages, seventy-one hundred needed replacement. Goldberg's powerful assessment of the Native land claims drew virtually no arguments. All that remained was the determination of the size of the settlement.

At the outset of the hearings early in 1968, the AFN had proposed that the aboriginal land claims be settled with 40 million acres and $500 million, an award that even its own supporters never thought they could get. On December 14, 1971, with the House voting 307–60 for a compromise package and the Senate whipping it through on a voice vote, Alaska Natives received 40 million acres of land and $962 million in cash. It was the largest land claim settlement in the history of the United States. (Subsequently, through a series of special-purpose amendments to include burial grounds and traditional sites, the total acreage grew to 44 million.)

Many whites in Alaska thought it was too much. Some Natives were displeased by the lack of specific commitments to the continuance of subsistence hunting and fishing rights. Charles Edwardsen's group, the Arctic Slope Native Association, felt it deserved mineral rights to Prudhoe Bay. But most observers agree that ten years earlier, before oil had been discovered, the Natives would have won nothing at all. Ten years later, they would not have received so much land, because the discovery of oil made it so valuable.

With a telephone call to the AFN meeting in Anchorage, President Nixon signed the bill into law on December 19, 1971. Twelve regional corporations within the state and one at-large corporation for nonresidents were designated to distribute the benefits. Anyone with one-fourth Native ancestry—just one Native grandparent—would receive shares, although nonresidents of the state would not be entitled to land. Alaska Natives had become Alaska capitalists.

CHAPTER 6 Birth of the Pipeline

Even in June, snow still covers the gentle peaks of Sugarloaf Mountain across the inlet at Valdez. Tour boats leave rippling wakes as they head out of the harbor carrying the season's early visitors to see the Columbia Glacier and the marine life that abounds in Prince William Sound. In town, a pleasant man behind the counter at the Hook, Line and Sinker bait shop swaps stories with idling customers as he fills fishing-reel spools with new line for the imminent first run of pink salmon.

A deep-port town of about three thousand people, Valdez sits at the head of Prince William Sound, which lies at the top of the Gulf of Alaska and is one of the most magnificent ocean waterways in the world. Humpback and killer whales play in the waters. Halibut and lingcod thrive in its depths. A great variety of birds, from bald eagles to tufted puffins, live on the cliffs of the many islands. Bears and goats live on the mountains that rise as high as 13,000 feet from the edge of the sound. Blueberries, cow parsnips, and herbs, among other edible plants, carpet the forest floor and alpine tundra.

The Chugach Eskimos, descendants of the Aleuts, first populated shores of the sound. They pursued orca whales in the sea, trapped salmon at the mouth of the streams, and hunted otters and seals. Whites first set foot in Alaska in 1741, when Vitus Bering led a Russian expedition that came ashore at Cordova at the extreme southeastern end of the sound. Bering's ship ran aground on the return trip and he died of scurvy, but those who made it back to Russia told of the abundant sea otters, setting off a rush in the fur trade that led to the near decimation of the species.

In 1778, Captain James Cook sailed up the sound trying to discover a northwest passage. In his published journals, he named the sound for the prince who subsequently became King William IV of England. The Spanish explorer Salvador Fidalgo, who also was searching for a northwest passage, stopped at the head of the sound in 1790. He gave Valdez its name, but Alaskans never used the Spanish pronunciation. They refer to it as Val-*deez,* and anyone calling it Val-*dez* is quickly stamped a stranger.

Valdez gained its first real notoriety at the turn of the century. When news of a big gold strike set off a stampede to the Yukon in 1897, more than twenty thousand people jammed into the little frontier town, turning it into a tent city. Valdez was not the shortest route to the Yukon, and the stampeders had to traverse 20 tortuous miles over a glacier, but it spared them interrogation at the Canadian border. Thirty-two churches vied for attention with a like number of bordellos until gold ran out and so did most of the people.

By the Depression, the town's population had sunk to five hundred, where it remained. Then on March 27, 1964, the worst earthquake ever to hit North America occurred, with Valdez 20 miles from the epicenter. Houses cracked, foundations sank, and massive tsunamis sent waves 50 feet high thundering across the waterfront. Thirty-two people died, and many residents abandoned their homes, the ruins of which are still visible today.

John Kelsey, lean and ruddy at seventy, is a Valdez pioneer who well remembers those days. He sits behind his desk at the Valdez Dock Company with a lake trout mounted on the wall behind him and talks of a lifelong commitment to the hardscrabble life in Valdez. A genuine Alaska boomer, he went to college at Stanford, he says, because even in tough times Valdez believed in education for its young. It helped that his family could afford tuition during the Depression.

For more than a half century, the Kelseys operated the Valdez Dock Company, one of the few profitable businesses in town. The giant sea waves that surged in after the earthquake demolished their docks, nearly swallowing John Kelsey with them. But the Kelseys stayed on and helped rebuild Valdez on a more solid site 4 miles away. By the time they put in their new dock, however, John and his brother, Bob, were $500,000 in debt.

When John heard of the gigantic oil discovery at Prudhoe Bay in 1968, the boomer spirit leapt within him. He knew the companies would have to ship the oil south out of some port in Alaska. Valdez, now reduced to a shabby trailer town with an unemployment rate of 40 percent, was in disarray, but it had a rare prize—the northernmost ice-free harbor in Alaska.

Prince William Sound is about the size of Chesapeake Bay. To reach the port of Valdez, ships must sail due north about 40 miles after coming into the sound from the Gulf of Alaska. The countless fjords and islands make it a breathtaking trip, and it has long been a popular route for cruise ships. But Kelsey knew that the location was more than pretty; it was strategic. Not only was Valdez closer to the oil fields than any other port in Alaska that stays ice-free during the winter, but a paved road, the Richardson Highway, connected it to Fairbanks, providing a direct overland link to Prudhoe Bay. Valdez had all the advantages in the contest for the commerce in the coming oil boom on the North Slope.

Yet far more influential people and interests had their own plans for shipping oil out of Prudhoe Bay. Secretary of the Interior Hickel wanted to extend the Alaska railroad from Fairbanks to Prudhoe and transport the oil in tank cars to Seward. Exxon, then Humble Oil, a major player in the planning, had early visions of using a northern water route. It spent $50 million to send a big ice-breaking tanker, the *Manhattan,* on a mission to determine whether escorted tankers could navigate to American markets through the often ice-clogged channels of the Arctic Ocean. After two bad trips, Humble gave up the idea.

Meanwhile, the Wilderness Society, Friends of the Earth, and other major environmental groups whose voices were rising in the nation opposed any shipping of oil by sea. They urged the government to force the oil companies to explore building a pipeline or a railroad across Canada to Chicago as a way of avoiding marine spills. But certain Native leaders opposed a pipeline. They feared the impact on caribou herds and fishing streams would doom their subsistence existence, and threatened to go to court to protect it.

In Kelsey's view, they were all off base. "Hickel's idea was a pie in the sky," Kelsey says. "All I could think of was a line of tank cars from Prudhoe Bay to Seward. The Canadian land route was bad because we wouldn't have had control of our own pipeline. And the Natives who were concerned about the caribou and fish were not aware that stiff environmental controls were certain to be imposed on the oil companies."

Kelsey wondered how a little dock owner from Valdez could get the ear of the oil companies. Then he remembered. When he had been mayor (Valdez has many ex-mayors, because the term is only one year), company officials nearly always came to the phone when told a mayor was calling. "So I got George Gilson, the mayor, and Bill Wyatt, who ran a motel, into my office and we made some calls," Kelsey says. "We tried to reach the Humble Oil president. We called Atlantic Richfield and British Petroleum. At first they wouldn't talk to us because they thought we were a couple of characters out of the night. We kept badgering and only asked that they listen to us. Finally, they gave in and told us about a study group working in Alaska." Kelsey tracked the study group to the Captain Cook Hotel in Anchorage. He called their suite and said the mayor of Valdez would like fifteen minutes of their time and was on his way.

The study team was just leaving the bar when the mayor and Kelsey arrived late in the afternoon. By coincidence, they stepped into the same elevator. "They looked at us with all the maps under our arms and said, 'You must be from Valdez,'" Kelsey says. "And they reminded us we had only fifteen minutes."

Inside the room, Kelsey spread the charts on the floor and traced a line from Prudhoe Bay to the nearest ice-free, deepwater port, which, of course, was Valdez. Then he pointed out the possible pipeline routes to Valdez that could maximize the use of federal land, enabling the companies to avoid costly land settlements. He started to get their attention. "The fifteen minutes dragged into three hours until someone on the team said, 'I'm hungry,' " Kelsey says. "They invited us out to dinner, and before the night was over I knew they were falling in love with our harbor."

In mid-February 1969, the three major Prudhoe companies—Atlantic Richfield, Humble Oil, and British Petroleum—announced without specifying a port that Prudhoe oil would be transported by pipeline and that the delivery system would be entirely in Alaska. Walter B. Parker, a member of the federal field committee comparing pipeline options, said the strong insistence upon West Coast delivery by Robert O. Anderson, the president of Atlantic Richfield, swung the decision. The field committee, appointed by President Lyndon Johnson, did not have veto power, but its chairman, Joseph Fitzgerald, had Johnson's ear and probably could have blocked the decision had the committee disagreed. Engineers were to determine the exact route and terminal, but the conduit was to be a 48-inch-diameter pipeline stretching for about 800 miles. Hydrostatic studies had determined that using a 48-inch-diameter pipeline would require the fewest pumping stations.

The first estimated cost of the pipeline was $900 million. *The New York Times* described it as "possibly the largest single private construction project and private capital investment in history."

Kelsey says he was informed six months in advance of the public announcement that the oil companies, now organized into a construction company called the Trans-Alaska Pipeline System (TAPS), had picked Valdez for the terminus. He was overjoyed. He saw the salvation of Valdez, as well as his own business. Then he showed his gratitude by giving the pipeline project one more big assist.

In 1968, the Internal Revenue Service had changed the tax laws to address the abuses of tax-exempt bonds. Some congressmen had been angered by the way southern states were floating these bonds to finance new factories that were luring textile firms from New England. Bond issues for docks and environmental safeguards such as those planned for the pipeline terminal would not be affected, however. The state of Alaska had the enabling legislation, and Governor William Egan agreed to float the bond issue until it was pointed out to him that with bonds issued by the state, the oil companies would save three or four points in interest—a gigantic break for a construction loan certain to exceed a billion dollars. The governor's advisers urged him to claim 50 percent of those savings for the state. The oil companies balked.

When Bill Arnold, a friend of Kelsey's and an attorney for the oil interests, told Kelsey of the impasse, Kelsey urged that they explore the possibility of using the city of Valdez to float the bonds. A few days later, Arnold returned and told Kelsey that Atlantic Richfield's Anderson was wary of turning to Valdez for help. He feared Governor Egan would retaliate by asking the legislature to raise taxes on oil. Acting on his own, Kelsey went to Juneau to visit Egan, a longtime friend and a fellow native of Valdez. When he left the governor's office, he had a handshake promise of no retaliation.

With the oil companies agreeing to guarantee the bonds, the little city of Valdez floated $1.5 billion of a $2 billion bond issue to build the pipeline—the largest tax-free industrial bond ever issued up to that time. The windfall for Valdez, which gets a 1 percent dividend each time the bond is refinanced, enabled the city to establish its own permanent fund for city improvements. Valdez became a boom city. Modern public buildings, including a community college campus and an attractive convention center, were built with oil revenues, transforming a faltering downtown into a bustling urban center. That attracted private investors to build first-class lodgings, restaurants, and boutiques.

The financing cleared the way for purchasing the 800 miles of 48-inch pipe that would provide the artery for the pumping system. Such a prospect should have brightened boardrooms across the American steel industry, but it didn't. No U.S. company had a plant to manufacture 48-inch pipe, and no American company could guarantee to build a special pipe mill in time to accommodate the crash schedule set by the oil firms. But the Japanese were ready to oblige. A $100 million order for the first shipment of pipe was placed with two Japanese mills, with delivery to start within five months.

This should have been a lesson for the American steel industry. But Peter DeMay, vice-president for project management on the pipeline, says that a few years later, when a need arose for fabricated steel chutes, no domestic steel companies bid on the order. The Japanese firms, aware that they had no competition, came in with overly high bids. DeMay got Edgar Speer, the chairman of U.S. Steel, on the phone. "Here I am trying to order some fabricated steel—and all it is is fabricated steel—and I can't get a bid out of you." A surprised Speer promised to have someone call. "Not good enough," DeMay said. "We don't have that time." DeMay said he sent his people down to the West Coast with drawings and U.S. Steel agreed to expedite the job. "The order was one hundred eighty million dollars, probably the biggest they ever had," DeMay said. "And it had to be forced upon them."

With Interior Secretary Hickel, the state's entire congressional delegation, and the governor of Alaska working hard to clear bureaucratic obstacles, the oil companies confidently started work on the pipeline early in 1970 without waiting for permits. TAPS bulldozed a haul road from above Fairbanks to Prudhoe, built an ice bridge across the Yukon, and began to move hundreds of men plus much equipment and material to construction camps along the route. But just as the oil companies were gearing up a construction spectacular to start oil flowing, other forces were organizing a spectacular attack to halt it.

Congress had passed the National Environmental Policy Act, the most sweeping of its kind ever, in 1969, almost coincidental with the decision to build a pipeline. Major environmental organizations, working in concert with the Center for Law and Social Policy in Washington, seized upon the act as an instrument to stop the oil companies from carving an 800-mile strip across Alaska's fast-disappearing wilderness. In this first major test of the newly formed Environmental Protection Agency (EPA), three lawsuits were filed in Washington's federal district court, starting on March 26, 1970. They were brought in the name of the Wilderness Society, the Friends of the Earth, and the Environmental Defense Fund.

The first suit charged that TAPS was asking excessive rights of way in violation of the Mineral Leasing Act of 1920 (which, while outdated, is still a handy weapon for all varieties of obstructionists). Next, the plaintiffs sought to join their suit with requests for an injunction filed by several Native villages that were refusing to let the pipeline go through their property. Then the plaintiffs filed an amended suit saying that the Department of the Interior had not complied with the impact statement required by the EPA.

This legal barrage produced court injunctions that stopped the project dead. For the next four years, loads of pipe and expensive machinery sat idle at Prudhoe, and many suppliers that had stocked materials there on speculation went broke. The battle raged in the courts, in Congress, in the media, in offices and bars. To many, the pipeline had become a national symbol of the destruction of the country's last bastions of wilderness.

In January 1971, an Interior Department staff report concluded that the pipeline would create unavoidable environmental damage but recommended that the line be built. Oil, it stated, was that crucial to the nation. With Native claims now settled, the oil companies thought the Interior Department's endorsement cleared the last big obstacle. But Hickel was no longer secretary of the interior. President Nixon had ousted him from office months before, after Hickel wrote him an impulsive letter complaining that the administration's Vietnam protest policy was polarizing American youth. Hickel's successor, Rogers C. B. Morton, promptly stated he was not bound by the finding of a report started under Hickel, and wanted to know more about the risk of piping hot oil across the permafrost. *The New York Times* applauded the position, proclaiming, "Hot oil is dangerous enough without hot haste."

In February, the Interior Department scheduled pipeline hearings, first in Washington and then in Anchorage. Alaskan leaders lined up a formidable array of engineers, government officials (including Hickel), and Native leaders to testify. But a few days before the hearings opened, a full-page ad appeared in *The New York Times* bearing the signature of David Brower, president of Friends of the Earth. It stated that two thousand oil spills had occurred in U.S. waters in 1966 alone, killing millions of birds. "If oil is pumped out of Prudhoe Bay and then shipped down the west coast," Brower warned, "we will eventually have an oil spill leading to the greatest kill of living things in history."

The Washington hearings became a showcase of unyielding positions. Pipeline advocates claimed Alaska oil was economically essential and the levels of damage and risk acceptable. Conservationists pointed to the industry's record of marine spills and predicted the destruction of Alaska's pristine wilderness if the interior were opened by pipeline construction. By the time the hearings moved to Anchorage, some interesting embellishments had been added to the debate. Arguing for the pipeline, Mayor Julian Rice of

Fairbanks testified, "God placed these things beneath the surface for a purpose. For us to say that we shouldn't use them is to be anti-God." And Senator Ted Stevens of Alaska was said to have told the Cordova fishermen who make their living fishing Prince William Sound, "Wernher von Braun, you know, the spaceman, assured me that all of the technology of the space program will be put into the doggone tankers and there will not be one drop of oil in Prince William Sound."

The Anchorage Times did a competent job of reporting the thrust of the arguments from the three hundred witnesses who asked to be heard in Anchorage. However, it, too, had its lapses into oil boosterism. "The fears about damage from oil spills are like the fears of Henny Penny when she ran to tell the king that the sky was falling," read one editorial.

The battle might well be raging today but for a gasoline shortage that gripped the American public in 1973. The official explanation for the crisis was that the OPEC cartel had imposed an oil embargo on the United States in retaliation for aiding Israel. Not everyone bought that explanation for the long lines at the gas stations, however. "I felt it was a contrived oil shortage then and I still think so today," says David Brower. "It was contrived to get congressional support for the pipeline."

Nevertheless, the nation's dependence on oil suddenly obliterated its budding sensitivity to the environment. Moving oil from Prudhoe Bay became a patriotic mission, and Congress took the cue. In several swift steps, it removed the obstacles stalling the pipeline. When the courts found for the environmentalists in their suit charging that the oil companies violated the 50-foot right-of-way limit in the Mineral Leasing Act of 1920, Congress simply amended the act to allow the secretary of the interior to increase the right of way temporarily "when any specific project merited it." And when it still appeared that new legal challenges under the Environmental Policy Act could delay the pipeline indefinitely, Congress introduced an amendment to the pipeline bill insulating its permits for construction from any new legal challenges. In the eyes of many senators, that was cutting the heart out of the EPA, and the vote was hung up on a 49–49 tie. But Vice-President Spiro Agnew came to the rescue, making sure he was there on July 17, 1973, to break the impasse.

Birth of the Pipeline

Before Congress ended its 1973 session, the House passed the pipeline bill 361–14 and the Senate passed it by a vote of 80–5. President Nixon signed it into law on November 16, and Interior Secretary Morton issued the long-awaited construction permit on January 23, 1974. Five years after it was conceived, work on the pipeline legally began. And the estimated cost had grown from $900,000 to $6 billion.

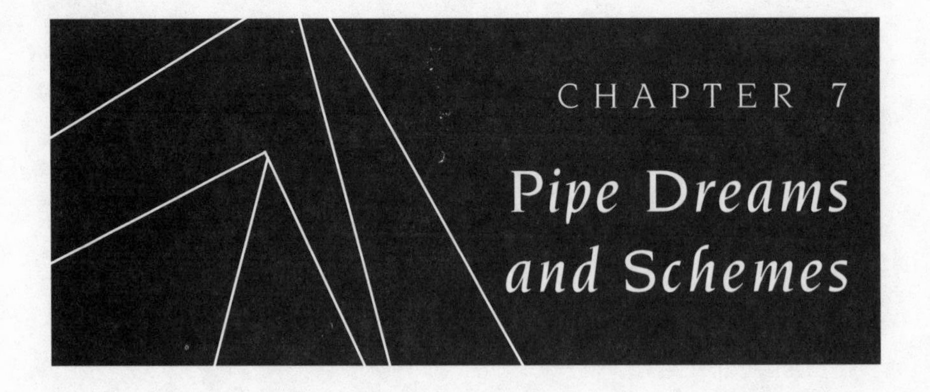

CHAPTER 7

Pipe Dreams and Schemes

IN FEBRUARY 1975, Susan Blomfield, young, pretty, and bored, was just out of high school with a career that was going nowhere. She was making $2.75 an hour in an Anchorage florist shop and her boss was on her case for taking too much time on floral bouquets. That's when she walked out and headed for the pipeline hiring hall in Fairbanks. "If I'm going to be hassled, at least I'll get a decent buck for it," she reasoned.

Susan had watched a steady stream of acquaintances head north to Fairbanks in pursuit of big dollars on the pipeline. Her brother, most of the Friday night gang at Chilkoot Charlie's, and even the cook at the Westmark Hotel lit out for the boom. The word was that they were hiring everybody in the rush to get oil flowing from Prudhoe Bay.

When she got to Fairbanks, Susan learned differently. "After signing up at every union labor hall and getting no response, I began to worry," she says. They were hiring all the craftspeople they could find, true. In fact, there were more crafts jobs than there were experienced craftspeople in all Alaska. But workers with no training went to the bottom of long waiting lists. With temperatures regularly dropping to thirty degrees below, Susan shivered with other unskilled workers every morning in lines that stretched three blocks from the hiring hall.

Susan became one more hapless figure in Fairbanks, an overrun city that appeared to be bent on reliving a notorious past. Founded in the gold rush at the turn of the century, Fairbanks started as a supply depot for the nearby mining camps. It soon became a center for gouging merchants, roving con men, and sleazy saloons offering whiskey and women, all of them intent on parting the miners from their money. When gold ran out in the early 1900s, Fairbanks turned respectable. A railroad linking it to Anchorage and the ice-free port of Seward made it a legitimate trade center for the vast interior region. The military built two major installations nearby, Fort Wainwright and Eielson Air Force Base, after World War II. And tourism flourished, complete with gold rush tours and steamboat rides on the Chena River.

The oil strike at Prudhoe Bay, 390 miles to the northwest, turned Fairbanks into an uncontrollable boomtown again. Located midway along the 800-mile pipeline corridor from Prudhoe to Valdez, Fairbanks was an ideal site for the warehouses and hiring halls. By 1975, its population had doubled in five years to twenty thousand as word spread that the oil companies were paying high wages to whoever was willing to work in cold Alaska.

By the time Susan arrived, the town was overflowing. The airport was strewn with the sleeping bodies of Outsiders who had risked their savings on a plane ticket north. One-room apartments, when they could be found, were renting at $600 a month. A hamburger that would have cost 75 cents in Tulsa cost $4 here. Residents talked derisively about the influx of Okies and Texans, but businesses happily took their money. Despite the high prices, shoppers backed up daily at grocery checkout counters, and the bars on Second Avenue (known as "Two Street" by the regulars) were awash with the laughter of pipeline workers in town for their week off. They bought drinks with $100 bills and casually polished the bar with the uncounted change as they regaled one another with their experiences on the job. Women nursing long, pink drinks sidled up to the spenders and joined in the hilarity.

Meanwhile, Susan continued to wait in line at the hiring hall, and her anxiety continued to grow. She noticed that people with pointed shoes and southern accents kept being hired ahead of her. She had heard there was a law requiring pipeline contractors to hire Alaska residents first, and she wondered what had become of it. But the men at the union hiring table only shrugged or gave her a sly wink when she asked.

The Trans-Alaska Pipeline Authorization Act passed by Congress in 1973 had indeed specified Alaska hiring preference, but it was regarded as a joke at the hiring halls. A racket in counterfeit Alaska residency cards blossomed right from the onset of pipeline construction, and the unions helped it grow. "You know the right bar or the right girl working the streets and you find out where you can get a counterfeit residency card," J. Randy Carr, a state labor department investigator, recalls. "Illegal cards were going for a hundred dollars a pop and no one in the system seemed to be concerned about it." Carr says he was in on a sting operation in which the state caught one seller with two hundred counterfeit cards in his possession. He was charged with "uttering a forged instrument," but the judge threw out the case. His reasoning? The defendant had represented the cards as forged, so he hadn't attempted to hoodwink anyone.

The unions' willingness to accept falsified residence cards was only one of the obstacles between Susan and a job. Another was her gender. The Trans-Alaska Pipeline Authorization Act also specified that the project conform to the government's affirmative action program, which banned discrimination by sex, yet Susan noticed that only men seemed to be getting jobs at the International Union of Operating Engineers table. When she complained, the man behind the table put it to her bluntly: "We have never hired women and don't expect to start doing so." The Operating Engineers were represented by Local 302, based in Seattle. "You can imagine how hard they enforced the Alaska hire statutes," Carr comments, "since nearly all their members came from Seattle."

Susan decided to hang in and fight. She got a job waiting tables at the Grubstake. There she met a cabdriver who told her to see Randy Parker, director of the Equal Employment Opportunity Center, and tell him about her experience with the Operating Engineers. Parker called the Operating Engineers and read them the equal employment opportunity provision in the pipeline act. The union decided that hiring Susan might be less troublesome than defending a discrimination suit. Besides, there were other ways to deal with the problem.

Susan was hired—to operate Cat dozer number 7, a job calculated to send her packing. Nervous and unsure whether she could handle the mammoth machine, she nevertheless entered the training program and kept at it after hours. To the surprise of everyone, including herself, she passed the test. She was ready to join the massive invasion of men and machines slicing an 800-mile corridor north to south across Alaska.

There was an urgency to work on the pipeline that was like the fervor of a military expedition. Crews landed by land, sea, and air all along the route. Then they raced to see who could move fastest. Nineteen camps, with random names like Happy Valley, Kennedy, and Sheep Creek, were built along the route. Each had a narrow, prefabricated dormitory, complete with dining hall, rec room, and first-aid station. Support planes, barges, and trucks brought in a steady stream of supplies, from the latest tools to thick sirloin steaks. Top-notch equipment and plentiful food came with the job.

A new construction consortium called Alyeska Pipeline Service Company had replaced the temporary TAPS operation. It formed a joint command for the eight oil companies that planned to ship oil from Prudhoe. The chief executive was Edward L. Patton, a Humble Oil refinery builder whose blunt, no-nonsense leadership style was as military as his name.

Design plans for the pipeline had been laid in Houston, where the best minds in arctic construction got together in command rooms at an office building once occupied by famed wildcatter Glenn McCarthy. Geotechnical engineers studied how to move hot oil across frozen ground. Seismologists worked with civil engineers to protect the pipeline from earthquake shocks. Naturalists mapped ways to accommodate the migration of caribou. Out of it evolved a pioneering design—a pipeline technology that even its critics agree exceeded all past technical horizons.

The project contractors were world famous. Bechtel Corporation, based in San Francisco, became the chief contractor for the main pipeline. Fluor Corporation of Los Angeles would build the twelve pump stations and the storage terminal on the bedrock across the inlet at Valdez. They, in turn, assigned construction work to more than a dozen subcontractors. Together, they unleashed an army of twenty-one thousand workers with orders to build the pipeline, pumping stations, and auxiliaries as fast as possible. Welding applicants got first consideration. It was estimated that seventy-one thousand welds would be required to string the pipeline from Prudhoe to Valdez. At least two thousand welders were hired, who in turn required two thousand helpers. About three thousand operators were needed to run the machines—fleets of trucks, bulldozers, and myriad pumps and generators. And that meant hiring thousands of additional helpers, because every machine operator had to have an oiler. Just about every conceivable craft—from carpenters to electricians to millwrights—was represented in the more than eighty job classifications on the project. The largest group were the support forces, ranging from laborers to bull cooks, who handled the jobs that machines could not or craftsmen would not. All were represented by seventeen separate unions, each with its own contract and jurisdiction.

Everything was geared for speed. Alyeska was prepared to accept higher construction costs any time the alternative meant delay. Each day lost meant the sacrifice of profits from 660,000 barrels of oil, the estimated daily flow at start-up. No one attempted to peg the precise figure; it was impressive enough to say that at $10 a barrel, oil companies would be giving up $6.6 million of income a day.

The consortium of oil companies wanted the pipeline completed in three years, and it paid heavily to ensure that goal. No contractor in the construction link had to worry about going belly-up. The contractors worked on a no-risk, cost-plus basis, which provided an incentive to meet the deadline without the concern of holding down charges. Well in advance of construction, Alyeska also got the seventeen international craft unions to sign agreements that no strikes, picketing, work stoppages, slowdowns, or other disruptive activity would be tolerated. In return, the unions won the most lavish worker wages in the United States.

The lowest paid were the laborers, many of them right out of high school, their $10.67 an hour nearly double the $5.80 standard wage for construction labor in 1974. And that was before overtime. The usual workweek at the outset was a "7-10"—seven days a week, ten hours a day—with every third week off. That grew to "7-12" as contractors pushed to meet deadlines. Because every hour after the first forty hours was at time and a half, with double time on Sundays and holidays, even the lowliest worker made better than $1,000 a week. The elite were the welders, or "pipeliners," most of them imported from Oklahoma and Texas, who averaged about $90,000 a year. The welders didn't have to meet the residency requirement because few Alaskans had the necessary skills. They did have to pass a stiff test before they were hired, and their welds on the job had to pass not only visual checks but X-rays as well.

In tales of goldbricking on the pipeline, welders emerged as the most imaginative. One worker claimed he saw a welder go a full week without making a weld. He pretended he was working but actually spent the time drawing maps with a torch on a piece of pipe, which he later sold as souvenirs for $500 apiece. Ducking work was easy, because without incentives to hold down costs, contractors often overhired. Jan Hansen, who became the state's workers' compensation hearing officer, worked on the pipeline as a timekeeper at Galbraith Lake, about 150 miles above the Arctic Circle. She says the place was so overstaffed that it was common to see people reading or sleeping on the job. One transferee from Atigun Pass in the Brooks Range sat and did nothing for a full week until someone noticed she was there. "And of the thirty-five people in our place, I was the only one who was ever an Alaska resident," she recalls.

However, nowhere was waste and greed more evident than in the domain of Teamsters Local 959. Led by Jesse L. Carr (no relation to J. Randy Carr), a union boss in the classic tough-guy mold, the union amassed a fortune in dues during the construction of the pipeline. At the height of its membership in 1975, the Teamsters banked about $1 million a week in its pension trust accounts at the National Bank of Alaska, according to news reports that the union did not refute. While some of that money came from Teamsters working in other occupations, Carr knew how to apply muscle to pipeline contractors. He won hundreds of featherbedding jobs by coercing deadline-pressured contractors to place more than the normal crew of workers on jobs in the name of safety or labor peace.

When decisions happened to go against the Teamsters, Carr was a master at circumventing the no-strike clause. He once demanded that Teamster-driven pilot cars accompany all trucks and that two Teamsters sit in each base ambulance, twenty-four hours a day. When Alyeska refused, he pulled all Teamsters off the job for emergency "safety" meetings, stopping work as effectively as any strike. The costly pampering of long-haul truckers became a pipeline joke. Their base pay was a comparatively modest $10.67 an hour, but Carr demanded and won a stipulation that they be paid for eighteen hours per day no matter how many hours they actually worked.

The Teamsters' attempts to claim control over anything on wheels got them into a running fight over job jurisdictions with the International Union of Operating Engineers, which also manned driving equipment. When a Teamster challenged, the Operating Engineer would walk off the job in protest, or vice versa. To keep the work going, Alyeska arbitrators would usually end up assigning both men to the job or simply rule for the Teamsters. The arbitrators reasoned that the Operating Engineers would be gone after the pipeline was built, but Alyeska would have to live with the Teamsters, who at that time controlled a third of the state's recruitable work force.

Carr's Teamsters Local 959 was an empire. Besides the pipeline workers, it was also the bargaining unit for the Anchorage Police Department, the Alaska Hospital and Medical Employees, the Alaska Roughnecks and Drillers, the Anchorage Independent Longshoremen, and Skagway Longshore Unit No. 1. "Teamsters Local 959 thinks it runs the state of Alaska . . . and that's about right," Alyeska chief Ed Patton said in a speech following a string of Teamster-inspired jurisdictional disputes in 1975. Peter DeMay recalls that Jesse Carr violated the terms of the no-strike clause and binding arbitration so often that Alyeska had to take him to court.

The *Los Angeles Times* and the *Anchorage Daily News* both assigned reporters to investigate stories that Carr provided Teamster jobs for hoodlums and ex-cons, particularly at the Fairbanks warehouse complex where police said a large percentage of the material and tools shipped for the pipeline project kept disappearing mysteriously. The *Daily News*'s series on the inner workings of Local 959, published in December 1975, won the fledgling paper a Pulitzer Prize. Rumors of Teamster ties to organized crime spread in the summer of 1976 when two Teamster shop stewards were abducted from the Fairbanks pipeline warehouse in mob style. Their bullet-riddled bodies were found weeks later, but police conducted only a cursory investigation. They were not about to get involved in the union's internal affairs.

Thievery, rape, and other violent crimes were rising steeply along the pipeline, and the state police often felt powerless to deal with them. (Gambling and prostitution, while rampant, were not considered major crimes.) A big part of the problem, the papers reported, was that pipeline authorities were late in passing on reports. "Alyeska is willing to accept a certain level of theft in order to buy labor peace," State Attorney General Avrum M. Gross was quoted as saying in a *Los Angeles Times* piece of November 18, 1975. "They'll do nothing to provoke the unions. They just want to finish that line. They've stayed about ten miles away from state law enforcement people."

The contractors took their cue from Alyeska, even condoning blatant cheating on state income taxes by an estimated one fifth of their employees. The young state had neither the manpower nor the methods to deal with tax chiselers, and contractors simply grinned when workers claimed as many as sixty deductions on their state W-4 forms.

The state's impotence in cracking down on crime on the pipeline became so apparent that a federal organized crime strike force from California moved in with the FBI to try to clean up. In 1976, after a two-year investigation, nine persons were indicted, including a former assistant U.S. attorney who earlier had been forced from office for mishandling an investigation of Jesse Carr. All nine either had their cases thrown out or were acquitted. Alaska's image as a haven for lawbreakers and crooks on the lam survived intact during the pipeline years. Carr's empire would fall apart later, but not because any arm of the law caught up with him.

There were more poignant social costs as well. Many of the thousands of workers who flocked to Alaska were grateful for the chance at boom wages. They came with the resolve to tough out the long workweeks and the interminable subzero winters to save money to buy houses or set up small businesses when they returned home. However, the absence was tough on families. No one kept count of the marriages broken up or the infidelities, but they were the talk of the pipeline camps. And as bloated as the salaries were, hordes left their jobs with no more money than when they started. Stories abounded of paychecks squandered on drinking, gambling, and trips to Hawaii or Las Vegas for a few days in the sun or at the casinos.

In Anchorage and Fairbanks bars, you could find drugs for sale out in the open. "People didn't do the stuff on the job or they would be fired," one pipeline worker says. "But on their week off, after working three weeks steady, they would come into town and just lose it." Valdez was the handiest recreation spot for pipeline families. "You would see children of pipeline workers buy their lunch with hundred-dollar bills," a community college president recalls. He says hustlers siphoned entire family budgets with trays of questionable gems, dubious art, and all kinds of gold items whose real value was difficult to appraise. Even after oil started flowing, some workers stayed on just trying to hold on to the carnival mood.

Job turnover was a nightmare for the bosses. More than fifty thousand people had to be hired to maintain the twenty-one-thousand-job level. Some quit as soon as they stepped out into the numbing winds at their first work site. Dropout by Natives was particularly high. A general statement promising minority hiring had been written into the pipeline bill. The hope was that pipeline income, combined with training in particular crafts, would equip Natives to participate in the general economy after the pipeline was finished. The Alyeska Pipeline Service Company worked out the specifics with the Alaska Federation of Natives and agreed on a goal of three thousand Native hires. That was exceeded by about twenty-one hundred before the pipeline was finished, but the good intentions were never realized.

More than half of all Natives hired worked for eight weeks or less. Most simply could not adapt to a white man's regimen of twelve-hour days and a camp life far removed from family. Many left their jobs to fish when the salmon ran and to hunt when the caribou migrated. An estimated 46 percent of the Natives who left were "involuntarily terminated." Some of these dismissals involved jobs that were eliminated, but most were dismissals for cause, most commonly for drunkenness, absence, or tardiness. "Time meant nothing to them," one supervisor complained. "They were used to subsistence life-styles and would not change their habits." Most of the Native workers who did stick it out came from the larger villages, such as

Kotzebue, Nome, and Dillingham, observed Charles Elder, Jr., a former vice-president of Alyeska. "They hung in there because they were being converted to a cash economy," he says. "They discovered what money was and what you could do with it."

Unfortunately, success bred cultural problems for some of the Natives who did hang in. With earnings ranging from $1,000 to $1,500 a week—and often more—many earned in one month what their entire families in the village earned in one year. (The median income for Native families in the villages was only $5,200 a year even in the mid-1970s, when they were already receiving dividends from their Native corporations.) The brief encounter with pipeline wages often caused estrangements at home. Some refused to return to the villages when the project ended, undermining the long-standing cultural tradition that called for family members to share their income, food, and energy for the benefit of all. This essentially left the elderly—the poorest and the weakest—to keep together the values, customs, and organizational structures of the community.

A good number who did return tell how their pipeline incomes enabled them to buy trailer homes, snowmobiles, and other luxuries. But they also developed an appetite for the cash economy, and that made them more aggressive in seeking changes to the old ways. And with some of those who returned, drugs came to the villages for the first time.

While Natives had the hardest problems adjusting to jobs on the pipeline, the nineteen hundred women hired on the project probably faced the worst obstacles. They had to overcome hiring discrimination, sexual harassment, and macho resentment.

Susan Blomfield knew all about it. Her first assignment was to Five-Mile Camp on the Yukon River near the Arctic Circle. Six hundred men were in camp and only eight women, including two prostitutes, two labor union staff workers, and four secretaries. Hookers operated openly, at a rate of $300 a customer, and Susan says she saw one man turn over his entire paycheck.

"A woman intent on doing the work for which she was hired found it hard to gain respect," Susan says. "We faced a constant threat from men. I carried a big crescent wrench wherever I went. Everyone sat around and got drunk after work. Men would come to the dorm, scratching on the doors and making propositions." The environment made it unpleasant for her to even ride the bus to the job site. The buses were driven by male crews who plastered the ceilings and inside panels with pictures of naked women and carried on a crude banter meant to be overheard.

Because she refused to sleep with her boss, Susan says, she often got the worst assignments. One day, with the windchill factor at one hundred degrees below zero, she was told to go out and make sure the Rollagon hadn't quit running. Rollagons are huge vehicles with broad tires that travel well in snow and do a minimum of damage to spongy tundra. They had to be kept running all night because starting them cold was impossible in the arctic winter. Susan waded atop 10-foot drifts to get to the machine. When she climbed into the cab, a gale blew the door shut and it was an hour before the wind died down enough for her to escape. Only the heat from the running engine had kept her alive.

After that, Susan says, they decided she should not go alone. "So they would tie me and a lovesick guy named Gilbert on a rope which was tied to the building. We would grope our way through the snow and he kept trying to kiss me all the way. I felt sorry for him. The men kept telling him that I really cared about him and that I would give in if he tried to kiss me."

Susan regularly made at least $1,000 a week. During one rush period when 160 pieces of equipment were needed at once, she and another operating engineer made $2,500 apiece for the week by working seven straight eighteen-hour days changing oil. Susan increased her savings by buying and reselling condos for fast profits on the steaming Fairbanks real estate market. She had amassed $100,000 by the time her pipeline job ended and it was time to move back to reality.

While working as a grade checker, Susan had met a dozer operator she liked. They got married, a fairly frequent ending for pipeline romances. (One bride clambered on top of the pipeline in a long white gown to exchange vows with her groom.) Susan now lives with her husband, Mike, in Fairbanks. Both closely monitor news about the pipeline they helped build.

But happy accounts of those years always seem to be matched by dark ones. Lori Keim, a pert, auburn-haired newspaper copy editor in her mid-thirties, saw her boyfriend, a bright student with an academic scholarship to Texas A&M, pour his income into drugs. "He was a smart and nice kid who just wanted to go to college," Lori says. "But his dad talked him out of it, said he'd be crazy to pass up the chance to earn big bucks. He could always go to college, he told him. So the boy got one of the thirty-dollar-an-hour jobs—he was a foreman working twelve-hour days. Before long he was drinking heavily and then deep into cocaine.

"Years later, when his brain was fully fried, he got a gun, held his dad hostage, and accused him of putting him on the road to ruin," Lori says. "Then he shot his father dead, and then himself. It was such a waste. That crazy, manic time took a toll on the people who worked on the pipeline and earned those incredible salaries."

The pipeline was completed in 1977. It made the three-year deadline despite a scary delay in 1976, when a subcontractor was caught trying to rush the project along by falsifying X-rayed welding reports. He had certified some with minor deficiencies as acceptable and failed to X-ray several other welds at all. Besides precious time, it cost Alyeska $50 million to dig up, redo, and recertify the welds to the satisfaction of the U.S. Department of Transportation and critics in Congress.

The final cost of the pipeline was $8 billion, nearly nine times the initial estimate. Unanticipated leaps in costs, plus steadily rising tanker shipping charges during the double-digit inflation of the 1970s, would have made the project nearly worthless had not the Arab oil embargo sent oil prices spiraling during the 1973–74 OPEC rebellion. University of Alaska economists estimated that the cost of shipping oil from Prudhoe Bay was now up to $10 a barrel, which was greater than the 1972 market value of crude oil delivered to refineries in the lower forty-eight. In fact, it took a second Mideast crisis—the 1979 Iranian revolution, which tripled oil export prices—to make the Prudhoe Bay development truly profitable for the oil companies and the state of Alaska.

Nevertheless, the pipeline is viewed as a technological marvel and has become a worldwide tourist attraction. It starts at pump station No. 1 on the tundra on the edge of the Arctic Ocean at Prudhoe Bay. On its journey south, it traverses the mighty Brooks Range at Atigun Pass, crosses the mile-wide Yukon River, and winds through cuts in spruce forests and river valleys until it reaches the Keystone

Canyon outside of Valdez. There, it leaves the bank of the Lowe River and rises like an encased stovepipe up the steep canyon cliffs on its way to the storage tanks at Valdez. Wherever the pipeline crosses rock it is buried in a trench, and it is elevated on stanchions over unstable permafrost that would melt if it made contact with the 140-degree oil flowing through the pipe. Slightly more than half of the pipeline is above ground. Helicopters and ground forces maintain round-the-clock security the length of the line.

On June 20, 1977, *The Anchorage Times* reported in a banner headline, "FIRST OIL FLOWS (After 8 Years, 4 Months, 10 Days)." The metered tap was opened, but a faulty valve caused an inferno four minutes after oil started to run. One man was killed and the line was shut for nine days. "Oil's There at Last," read the headline on July 29. The first load of oil had reached the terminal and was on its way. On August 1, the tanker *ARCO* sailed out of Valdez with the first shipment and headed for California. The twelve pump stations began sending about 1.2 million barrels of oil a day from Prudhoe to Valdez. Soon, the pipeline would provide about 25 percent of the nation's oil needs, and fund about 85 percent of the state's budget.

Alaska became wealthy overnight—in fact, wealthier than anyone imagined. When world oil prices tripled, revenues poured into the state treasury by the billions. Poor Alaska had finally struck it rich.

CHAPTER 8
Days of Milk and Barley

IT IS LATE SUMMER. I am in my third year in Alaska, and the state is changing fast. I am about 65 miles out of Anchorage on the other side of Cook Inlet, and I have just run out of tarmac road. A dusty sign on the gravel says "Point MacKenzie," and the dirt road heads directly into a heavy cottonwood, spruce, and birch forest. Sleeping Lady, the crowning contour of Mount Susitna, is on the horizon.

This looks like good moose and bear country. For miles there is no sign of life, yet the deeper I drive into the woods, the clearer it becomes that someone intended this road to leave a permanent mark upon the land. I see torn-up clearings with mounds of dirt and stumps on either side of the road, evidence that man has tried to rearrange nature. And here and there in the wilderness I make out deteriorating barns, rusting farm machinery, and thousands of acres of fallow fields that are being reclaimed by weeds and scrub.

Far down the rutted road, I find a survivor, a sixtyish man who peers at me from under the hood of an old truck that sits in a muddy barnyard. Ducks, geese, and sheep scatter as I approach.

"Can you talk while you're working?" I ask.

"That's okay, I can take a break," he says. He seems happy to have a visitor.

He introduces himself as Harvey Baskin. The desolation I saw along the way, he explains, is the remains of Alaska's attempt to set up a dairy business. He hears the state has poured at least $100 million into its grandiose agriculture project, of which dairy farming is the heart, but the only thing it accomplished was to protect the jobs of bureaucrats. Harvey says he is but one of two of the original dozen or so homestead farmers whose farms haven't been repossessed. He, too, is behind on his loan payments and he doesn't know how much longer he can hold on.

The story of how Alaska tried to develop a dairy industry in the subarctic frontier involves adventurous men and women like Harvey Baskin. Their saga belongs alongside that of the miners and trappers, breeds who pride themselves on endurance. But the story also involves foolish decisions at high levels that created a costly and environmentally destructive boondoggle.

The state did not know what to do with the gusher of money that poured in after oil began to flow from Prudhoe Bay in 1977. Alaskans had expected the royalties would ease the lean state budget and flesh out capital expenditure. But even the most optimistic projections did not foresee the world events that would send oil prices into orbit. By 1981, oil had reached a high of $34 a barrel, an increase of more than 1,000 percent in thirteen years. As owner of the Prudhoe Bay oil fields, the state collected about 12 percent in taxes and royalties from the sale of every barrel piped to Valdez. During the Iranian crisis alone, Alaska's revenues tripled to $4.5 billion.

A few years before, the annual state budget had amounted to only $368 million, supplemented heavily by the federal dole. In 1969, the sale of Prudhoe oil leases had generated a modest windfall of $900 million. But that was a onetime bonus, quickly gobbled up by the long-starved needs of schools, hospitals, social services, and transportation.

Dealing with a steady stream of wealth was a task for which Alaska was not prepared, and a free-for-all promptly developed among the governor, the house members, and the twenty senators. The legislature—an assortment of fishermen, miners, lumberjacks, lawyers, real estate agents, small businessmen, and sourdoughs—began to act like a frontier version of Tammany Hall. The legislative committee chairmen met in secret to divide the largess, circumventing the normal budget process. Indeed, the revenue system was so overloaded that no one really knew how much money was coming in or being spent. Administrators and lawmakers accused one another of doctoring figures to their own advantage. The press made much of the disparities, some as high as $100 million, between revenue reports put out by the Republican administration and those of the Democrat-controlled legislature.

The oil industry watched quietly from the sidelines. The battle nicely diverted attention from proposals to adjust depletion tax allowances in the state's favor. They had less chance of passing as long as the legislature was in disarray.

Into this chaos stepped Russell Meekins, Jr., a thirty-year-old Young Turk Democrat from Anchorage who was the son of an auto dealer. He wooed other disenchanted Democratic legislators who had been ignored in the pie-splitting and quietly forged deals with the Republican minority, who had been utterly excluded. By the middle of the 1981 legislative session, Meekins had secretly formed an insurgent coalition. Seizing a moment when the speaker's chair was temporarily unoccupied, Meekins moved in, called for a motion to remove the current speaker, and declared the motion passed before the old guard, meeting in caucus, knew what had happened. His band of insurgents responded on cue.

In the uproar that ensued, shouts of "kangaroo court" and "banana republicanism" echoed in the chambers. The deposed Democratic leaders were certain the courts would restore their control. But incredibly, the courts upheld the coup, ruling in effect that they wanted no part in participating in the internal organization of another branch of government.

Meekins then supported a deal that everyone could understand. One third of the surplus from the oil revenues would go to the governor, one third to the senate, and one third to the house. It was understood that the governor would not question how the legislature spent its allocation and the legislature would not question how the governor disposed of his—an open invitation to conflict-of-interest governance.

No one took advantage of the deal more audaciously than Republican governor Jay Hammond. Elected to a second term in 1978, Hammond had barely survived a bitter primary, which he won by ninety-eight votes. A former hunting and fishing guide, he had been hammered hard by his critics—his primary opponent, ex-governor Hickel, foremost among them—for being a "no-growth" environmentalist. Hammond needed to blunt the charges, and an idea came to him. Why not use oil revenues to create an agriculture industry in Alaska? That would put a stop to the harping for more development, and it would also bring into the state a clean, green industry, consistent with Hammond's environmental leanings.

The governor called in his top strategist, Bob Palmer, and told him to make it happen. Palmer turned to the University of Alaska at Fairbanks for help. In time, Palmer had what he wanted: an authoritative-sounding feasibility study backing Hammond's dream. The report found "a general consensus that production of feed grain (principally barley) for export and to supply in-state livestock enterprises would be a logical thrust for development." The professors predicted that agriculture could make Alaska a national breadbasket, like the Midwest. The administration predicted that the state would be self-sufficient in eggs and milk by 1990.

No one questioned whether the finding might be slightly self-serving. The proposed barley farms would be virtually in the university's backyard and held out the ongoing potential for lucrative, state-funded consulting contracts. Nor did anyone ask how remote and undeveloped Alaska could ever get a competitive edge in the international barley market, or if, in fact, such a market existed. Only state senator Victor Fischer, founder of Alaska's Institute of Social and Economic Research, a think tank, came close to challenging the governor's plans. Noting that the state was preparing to bankroll farmers with millions of dollars in credit, he calculated it would cost an estimated $10 in state dollars to produce every $1 worth of agricultural goods. He politely wondered whether this might not be too high a ratio of subsidy. The governor told Fischer not to worry. After all, hadn't university experts endorsed the plan?

Hammond's great leap forward into agriculture designated 84,000 acres in the Delta area, deep in the interior just southeast of Fairbanks, as Alaska's barley-producing capital. About 70,000 of those acres were still in spruce forest and had to be cleared, but the program was in high gear by 1982. Alaskans of all sorts entered the lottery for barley-farm land. Easy loan money from the state's Agricultural Revolving Loan Fund put them in business, and promises of an elaborate transportation system to link them to world markets in Korea, Japan, and Asian markets beyond further buoyed their hopes.

The state proceeded confidently. It bought $940,000 worth of railroad grain cars to carry Delta barley to Seward, where the footings for an $8 million grain terminal were being erected in the harbor. Although Delta was 60 miles away from the nearest railroad track, it was simply understood that adequate linkage would be provided once barley fields started ripening. Meanwhile, barley fever had so seized the state that Valdez, closer to Delta and accessible by the Richardson Highway, built a rival grain terminal in its harbor. This one initially cost $15 million, financed by municipal bonds. Not a bushel of barley had been ordered by any international buyer up to this point, but the governor and his advisers were confident that a great new industry was on its way.

It had been intended from the outset that Alaskans themselves would be large consumers of Alaska barley—or, rather, their cows would be. A few small, family-owned dairy farms had existed for years in the Matanuska Valley outside of Anchorage, but this was not the market Hammond had in mind. He wanted large-scale dairy farms with hundreds of cows. Their needs would assure a steady in-state market for barley, and, in turn, a steady supply of milk—or so the state promised. True to its word, the administration set aside 33,000 heavily forested acres on Point MacKenzie, excellent moose and bear habitat, for dairy farming. It held a second lottery to give away about 14,000 acres to would-be dairy farmers, who would become eligible for a million-dollar line of credit from the state. No prior farming experience was necessary. The state would issue each winner a set of instructions on how to establish a dairy farm.

Harvey Baskin didn't win, but his wife and daughter hit it big and happily turned over their acreage to him with the understanding that they would not have to live on the farm. A retired chief master sergeant, Baskin had high hopes of carving out a second career in the wilderness. He dug into his savings to meet the 25 percent collateral that made him eligible for the million-dollar line of credit.

"The state gave us an opportunity you couldn't get anywhere else in the world," he says. Further, he claims, he was assured that the state-financed Matanuska Maid creamery would buy all his milk at good prices. A state-financed slaughterhouse, Mt. McKinley Meats, would buy calves and other expendable livestock for market. And good-quality grain was promised from the state-subsidized farms in Delta; he would buy it through the state-subsidized supply store in Palmer.

Baskin says his doubts about the state's competence to administer such an ambitious agricultural project started the day he saw his land at Point MacKenzie. "All there was was a gravel road, some stakes the state had driven in, and no state employee to show you anything," Baskin says. "The land was so wet you couldn't get a tractor on it. Maybe a hundred fifty of those five hundred acres were farmable. And I later found out we were one of the few farms that ever put up honest collateral for our loan. Hell, I learned some got their farms without putting up a penny. They listed nonexistent gravel pits as collateral. The state never checked."

Baskin reasoned that feeding an anticipated herd of one hundred cows by hand did not make sense, so he brought in a dairy expert to automate the process. "The state inspectors hopped on me for changing the design," Baskin recalls. "They said it was a violation of my loan agreement, and they threatened to cut off the money. It made no difference that I told them this was standard design in Minnesota. I had to redo the alley in front of the stalls to please the local bureaucrats. So I'm feeding the cows by wheelbarrow."

Baskin lived by himself in a small trailer on the edge of a forest for eighteen months while he cleared the usable land. Operating his own bulldozer, putting in fifteen-hour days, and grudgingly conforming to the state inspector's nit-picking rules, he met the state's thirty-six-month deadline and had his dairy farm operating by the fall of 1985. But Baskin's herd of 102 holsteins had barely gotten to know their stalls when the Matanuska Maid creamery, the farm's main source of income, filed for bankruptcy.

The state creamery had been paying Alaska farmers a generous $23 per hundred pounds of milk at the outset of the Point MacKenzie program. However, this price became highly uncompetitive in Alaska's new oil-based economy. Increased demands generated by a growing and increasingly affluent population made it more profitable to ship in large quantities of fresh goods in huge, refrigerated container ships with fast roll-on, roll-off service. Ironically, it was Sun Oil, an eastern firm that was not among the Prudhoe developers, that exploited this lucrative market. Its new 790-foot container ship *The Great Land* started service to Alaska in 1976 to deliver supplies to the pipeline, but quickly switched to hauling food in by truck-trailer. Eliminating loading and unloading by crews of longshoremen not only made service cheaper but allowed a guarantee of two-and-a-half-day delivery for perishables. By the early 1980s, Alaskans could buy milk cheaper shipped in from Seattle, some 2,000 land miles away, than they could from the local Matanuska Maid creamery.

Unable to adjust to the realities, the creamery nevertheless tried to maintain its promised price to Alaska farmers by borrowing on its assets, which essentially were the state's pockets. When it could borrow no more, it collapsed—$3 million in debt. The whole Point MacKenzie dairy project was now in deep trouble. The price of milk at the creamery dropped to $20 per hundred pounds. Angry farmers filed lawsuits, claiming the state had concealed the creamery's precarious financial position and, in effect, had misrepresented the economic potential of dairy farming at Point MacKenzie. When William Sheffield, a Democrat, succeeded Hammond as governor in 1982, he froze all plans for further agricultural subsidies pending a complete review. That further distressed the farmers who had cleared the land and had already started milking cattle.

Sande Wright, a contractor who grew up on a Montana cattle ranch, was the first dairy homesteader to go under. In his late thirties, with a thick crop of curly reddish hair and the body of an athlete, Wright has an easy smile, but it develops a nervous twitch when he describes his dairy adventure in Alaska.

Wright said he invested $500,000 of his own money—"everything I ever made in life"—to acquire and clear 640 acres of Point MacKenzie land. With the help of a million dollars borrowed from the Agricultural Revolving Loan Fund, Wright had modern dairy barns on the site and started milking a herd in excess of one hundred cows by the fall of 1985. He also brought in a double-wide mobile home from Anchorage and erected a handsome knotty pine front, establishing the finest residence in Point MacKenzie. It was just down the road from the new elementary school, which was part of the state's grand plan to develop a booming agricultural community.

Wright's vision of a good life on the edge of the wilderness began to blur even before the creamery went under. He discovered that the dairy farming techniques he learned from his father did not work in Alaska. For one thing, he found he could not afford to raise his own calves. "Because of the feed expense, it was costing us twelve hundred dollars to raise a heifer from the time she was dropped," Wright says. "We could buy really top heifers out of Canada for a thousand dollars or under, so we quit raising calves. We sold some and gave others away. And I would say half of the bulls we ended up knocking on the head. There was no market for them. We told the dog mushers on Knik Road to come and get them."

Wright tried to keep his farm going by pumping in money he earned with his earth-moving equipment, mainly his five bulldozers. When the creamery failed and the price of milk dropped, losses mounted to $15,000 a month. He now had no chance of meeting the $12,000 monthly payments on his $1 million loan. Wright filed for bankruptcy to stay the repossession of his farm, but there was no escape. "They dragged me into court four times trying to get relief from the bankruptcy stays so they could repossess the cattle," he says. "On the last try, I handed them the keys. They ended up butchering all of the cattle . . . told the slaughterhouse to take them all."

In the end, Wright lost not only his farm but his contracting equipment. His struggle to meet his loan payments contributed to the breakup of his marriage. And he owes the Internal Revenue Service $100,000 in unpaid withholding and Social Security deductions for his farm help. He cannot even start anew as a contractor because no company will bond a person in such fiscal distress.

The entire Point MacKenzie dairy development was soon plunged into financial jeopardy. In all, the state loaned about $20 million to the farmers. Few made even a single payment before their farms were repossessed. Outside of Harvey Baskin, who looks to his lawsuit against the state for salvation, only one other homestead dairy farmer remained in business by the beginning of 1991. That farmer was an Anchorage physician who managed to ward off repossession mainly because the state wiped out a $1.4 million loan in order to keep some Alaska milk flowing into the struggling state creamery. Even this sweetheart deal collapsed by midsummer. With unforgiven loans totaling $2.4 million, the physician walked away from his debts and gave the farm back to the state, leaving Baskin the sole surviving dairy farmer by 1992.

While advertising that it sells Alaska milk, Matanuska Maid stays afloat by purchasing most of its milk from Washington State. "You'd have to put it under a microscope to find the Alaska milk," Baskin says of the current product. Meanwhile, a court order has put a freeze on all of the repossessed farms, and they are deteriorating rapidly on this once unspoiled land. Vandals have torn Wright's handsome house to pieces. Windows have been smashed, appliances mashed into rusting parts, and a fire believed started by squatters has burned a hole through the roof. A reporter who walked into the stables a year after Wright departed found piles of manure still covering the floor and the skeleton of a calf in a pen where it had died. The elementary school where twenty-nine children once studied had been torn down. All that is left is a macadam lot littered with debris, a rusting swing stand, and a seesaw.

It was inevitable that the failure of the dairy farms would hasten the demise of Alaska's barley program in Delta. However, the barley farmers were in trouble long before the creamery failure knocked the stanchions from under the milk farmers. The eighteen grain cars bought by the Alaska railroad to haul barley arrived in 1982. They were painted a royal blue and bore the legend "Alaska Agriculture Serving Alaska and the World." However, they never delivered a grain of barley to either Alaska or any other part of the world.

Not only did Alaska fail to find an international barley market, but its sparse and uneven crop could not compete with grain from the Midwest in the lower forty-eight. Governor Sheffield halted construction of the $8 million grain terminal at Seward in 1982 when it was half built. Seward sold some of the I beams and bin-loaders to a Fairbanks dog food manufacturer in 1989, but recovered only a fraction of the investment. The rival grain terminal at Valdez fared worse. It was completed, costing the city an estimated $30 million when interest costs were computed. Never having housed so much as a grain of barley, it still stands in the harbor.

Like the dairymen, the barley farmers never got their production up to the speed envisioned by the state. Bulldozers readily cleared the trees from the 70,000 acres of land designated for barley in Delta, but it was not so easy to cope with myriad other obstacles. Voracious grasshoppers, extended droughts, dropping prices, and even roving bison beset the barley farmers. The Alaska bison, the largest native land mammal in North America, is descended from buffaloes transplanted to Alaska from Wyoming in 1928 to add still more variety to big-game hunting in Alaska. However, roaming across fields in groups of fifty or more, the bison became major pests for the barley farmers. Rather than extend the shooting season or relocate the herds, the state seeded a 2,000-acre barley field across the highway especially for the bison. It took the ton-sized animals all of a month to trample their own barley to the ground and discover better pasture in the farmers' fields.

In 1984, the most active barley farming year, only 16,700 acres in the 84,000-acre Delta region project were harvested. The planting dropped steadily after that, sinking to fewer than 5,000 acres by 1988. Matanuska Maid Farm Supply, the biggest customer in the state, had to start importing barley from the lower forty-eight to support its modest market of horse ranches and family farms. Most barley farmers started to make money only after they stopped planting. They found they could let the fields lie fallow and collect $37 an acre from the U.S. Agricultural Stabilization and Conservation Service under a subsidy program created by Congress in 1985.

Of the more than $40 million Alaska loaned to farmers since Governor Hammond set out to develop an agriculture industry in Alaska, $30 million is still delinquent. Counting the money spent on grain terminals, slaughterhouses, the creamery, and railroad cars, the state's expenditures exceed $100 million. And Alaska is neither agriculturally self-sufficient nor close to it. The only beneficiaries are the well-paid workers in the state department of agriculture, a cadre of about forty executives and support staff—about the same number as when the farm program started—most of them housed in a handsome office building in Palmer. As a bitter Harvey Baskin puts it, "The administration didn't know a damn thing about dairy farming. The program was administered by salary people in career-type jobs. It didn't make a damn bit of difference whether they were overseeing one or fifty dairy farms. They got the same amount of pay."

"We were going to make Alaska an agricultural state," Jay Hammond reflected after leaving office. "Frankly, had we not had oil revenues of the magnitude at the time, it would have been a gamble I never would have considered."

CHAPTER 9
Season of Gold

FOR A SHORT PERIOD in September, Alaskans enjoy a spectacular season before summer moves into winter. The changing colors of the birch and aspen trees turn mountain foothills into unending vistas of gold. For a moment, Alaskans bask in the glory and try to ignore the shortening days and the snow that dusts the tops of the mountain ranges. These are precious days, and Alaskans make the most of them while they last.

A similar euphoria overcame the governor and legislature of the state of Alaska in 1982. Oil revenues were flowing in so copiously that the state found itself with $4.5 billion in the till, of which it needed only $1 billion to balance the budget. During the debate over the pipeline, some Alaskans had wondered about the downside of oil discovery. How much progress was good for the state? Would rapid development ruin the pristine land? Would the influx of Outsiders change the unique character of Alaska? But with oil revenues flowing in, the downside was forgotten. Instead, there was much talk of Alaska's becoming a "superstate." It now had the means to create an environment that would attract industry, bring lasting employment, and end the instability of a boom-or-bust economy.

Capital gap. That, nearly everyone agreed, had been the biggest obstacle holding back Alaska. "There is no capital for borrowing in Alaska" is the refrain Robert E. LeResche, a Dartmouth-educated cabinet member in several state administrations, recalls from the late 1970s and early 1980s. " 'Don't underestimate the strength of Alaska entrepreneurs,' people were saying. 'Just give them a chance and they will prove that Alaska is ripe for business that will create jobs that will last long after oil is gone.' "

Having rolled the dice on Alaskan agriculture, Governor Hammond wagered a substantial bankroll of oil money on Alaskan capitalism. "The best way to keep the wealth in Alaska is to sponsor research to develop technology and innovation to advance renewable resources," reads a press release issued from the governor's office on December 15, 1978. It announced that Alaska was henceforth placing 5 percent of its oil wealth in a fund that would sponsor research to "identify new products, markets, and technologies for renewable resource industries." On this high promise was born the Alaska Renewable Resources Corporation (ARRC). It was to become a deep well for venture capital, funding projects that normal financing institutions would not handle.

The state gave the new agency an infusion of more than $40 million, including $15 million earmarked for commercial fish processing assistance. The governor appointed three board members, at cabinet-level salaries of $75,000 each, and sent them forth to ensure the future viability of Alaska. They were led by William Spear, a lawyer turned lapel-pin designer, described by some as a dreamer. Spear and his colleagues had great visions of Alaska's future, but, as audits would show, a decided blindness to established procedures for investment activity.

One of the board's first acts was to bring in a Boston consultant, at a $38,000 fee, to set up skills tests to help screen the qualifications of loan applicants. He devised a three-stage aptitude test. Would-be venture capitalists were judged on how well they tossed rings onto a pegboard on the floor, how precisely they stacked building blocks (some of which had rounded edges), and how wisely they selected equipment to ensure survival in the wilderness.

Word of the tests created an uproar in the legislature. State senators castigated ARRC for playing "kiddie games" while ignoring important qualities such as work experience and credit ratings. The tests were soon abandoned, but controversy over how ARRC doled out the state's substantial lending resources continued to simmer.

Doubts about the agency's credibility grew with its first major loan, a $330,000 grant to a freshly organized venture company whose principals included three people who had recently been members of Governor Hammond's administrative staff. Known as Tepa, their company proposed to build a plant to develop a chemical method for turning fish scraps into commercial protein.

With an announcement heralding the project as possibly having global impact, ARRC did considerably more than a traditional lender would to help the company succeed. It provided a two-year moratorium on loan repayments and required no interest payments until the third year. While the much-publicized purpose of ARRC was to create jobs and products in Alaska, Tepa decided to build its plant in the friendlier climate of Coos Bay, Oregon. The company told ARRC that the out-of-state site was necessary because no year-round supply of fish wastes was available in Alaska. It was hard to see how this was going to help Alaska employment or why the ARRC people bought the sales talk, but no one pressed the issue.

After several infusions of state money, it soon became plain that the venture needed far more than a steady supply of fish waste. The plant never overcame cost overruns and start-up problems. It went bankrupt without ever producing a product or making a payment. Auditors listed the state's loss at $410,581. "There were a lot of politically influenced interventions in which we were told to disregard the parameters," Wayne Littleton, ARRC's first executive director, concedes.

ARRC's investment judgment hit bottom in the case of the Salamatof Seafoods, a fish processing plant in Kenai, which sits on a point midway up Cook Inlet. The agency poured $4,961,000 into the plant but could not save it from bankruptcy. After the business lost $2 million during the 1980 fishing season, ARRC moved in to salvage what it could of its investment. It ended up buying out the owner, Tom Waterer, whom Littleton describes as "quite a charmer." Waterer happily accepted a $700,000 loan forgiveness in exchange for 50 percent of his shares and left the state to put the company into bankruptcy and deal with dozens of angry, unpaid fishermen.

A state audit in 1981 found that with the exception of one plant, all of ARRC's numerous investments in seafood facilities had failed or were on the edge. The agency's loan losses exceeded $15 million. Poor management, inferior salmon runs, Japanese competition, and a botulism scare were among its more credible excuses for the dismal record. In one case, however, a sizable investment loss was blamed on the stubborn religious beliefs of a fishing fleet operator who refused to work on Saturdays.

ARRC also invested heavily in nonmarine ventures, but they fared no better. The agency sank $200,000 into a mushroom farm that grew excellent mushrooms but failed to turn a profit because the farmer insisted on delivering the orders himself, socializing excessively at cafés along the route. It put $150,000 into a fox farm at North Pole, a borough near Fairbanks, but disease killed the entire breeding stock of two hundred foxes and, with them, any likelihood of a return on the investment. It lost about $3 million in a politically mandated loan to a sawmill operator who was going to build wood-fired powerhouses in Haines. The struggling little town hadn't developed much of an economy beyond the tourism it drew as gateway to the historic Chilkoot Pass of gold rush fame. News of the state's investment raised high hopes for a year-round industry, but the sawmill went bankrupt before it could complete even one powerhouse.

Agency directors had a hard-luck story for nearly every failure, but state auditors found a pattern to the series of loan defaults: the agency dispersed money blithely (one man got a grant for a proposed dog-powered washing machine), and often without even checking previous business failures or demanding specific project plans.

When he succeeded Jay Hammond as governor in 1982, William Sheffield summoned Robert E. LeResche and put him on the ARRC board with instructions to "straighten out the terrible mess." LeResche and two other new board members phased out the agency. The state replaced it with the Alaska Resources Corporation (ARC), which tried to operate with more specific criteria and accountability for venture capital investments until it, too, was phased out by Sheffield. "The ARRC experience proved it was impossible for a public official to run a legitimate banking operation," LeResche says. "If you try to apply sound lending practices, people say to you, 'Come on, it's not your money. It's the state's. I want my share.' "

The experience also left LeResche unimpressed with the Alaska entrepreneur of the eighties. "These people learned how to squeeze the state tit—all in the name of free enterprise."

If the venture capitalism undertaking proved disillusioning, however, the hope of investing in self-sufficiency for Alaska lived on, fueled by desire on the part of the governor and legislative leaders to leave their mark on history. While ARRC and the Agricultural Revolving Loan Fund were lending millions to sea and land projects, timber cutters and commercial fishermen pressured the congress for their own funding agency. And so the Commercial Fishing and Agriculture Bank (CFAB) was formed as a kind of adjunct to the Farm Credit Administration, a federal loan agency that was broadened to include fisheries in the 1970s.

CFAB was to serve as a cooperative bank for Alaska residents, although what constituted residency was never defined. It would operate much like an independent bank, with professional standards, except that it needed $32 million from the state to get started. This would be deposited with the Farm Credit Administration via the Spokane (Washington) Bank for Cooperatives, thereby qualifying CFAB to acquire more lending funds from the federal system. Those who feared this might prove yet another way to siphon off money from the oil-rich state treasury were assured that the state was totally protected. Alaska would receive class C stock for its $32 million start-up investment with a guarantee that the state could cash in the stock for full reimbursement in the year 2000.

The cooperative bank was launched with "exuberant myopia," in the words of one director. To establish its credibility as a professional lending institution, the management team decided it had to exude professionalism. It spent $246,835 furnishing its Anchorage office. The motif was teak—teak executive desks, teak coffee tables, teak bookcases, and a teak cabinet with a refrigerator and a stock of liquor for the president's office. Commissioned paintings decorated the staff offices, and a large soapstone carving of a land otter with a crab sat atop a rosewood conference table that itself cost $16,220.

The fishermen, processors, and lone potato farmer who made up the board were financial amateurs. To broaden the agency's vision, they decided to take their spouses on an expense-paid fisheries tour of Denmark, England, Rome, and Paris. Next came an agricultural tour through Scandinavia and an administrative trip to London. Other trips were scheduled for Japan, Munich, and Hong Kong.

Lawmakers demanded an audit before CFAB had been in business even a year. The auditors found thousands of additional dollars spent on a private apartment maintained in Juneau, lavish dinners in Anchorage's best restaurants, and bonuses for staff and directors. "A significant number of the expenditures may be inappropriate for an entity subsidized by public funds," the audit noted. CFAB responded that it was a private business operating under standards of private business.

That was too much for state senator Arliss Sturgulewski of Anchorage, a down-to-earth woman not given to outrage. Expressing strong concern over "questionable management expenses," she asked CFAB how it intended to approach repayment of the state's investment. The response of CFAB's lusty chairwoman, Roseleen "Snooks" Moore, who regularly fished her own boat in dangerous False Pass in the Aleutians, was uncharacteristically subdued. Repayment, she wrote, is "a multi-faceted problem that must be approached carefully."

It turned out that CFAB was in financial trouble almost as soon as it started. It was no more judicious in its lending procedures than it was in its administrative expenditures. It had virtually shoveled out $100 million in loans by 1984, and 35 percent of those were "nonperforming."

The cooperative lost millions financing crab boats at a time when Alaska crabs were disappearing from overfishing. "People were standing in line borrowing money for crab boats," Forest J. Paulson, then CFAB's chief executive, recalls. "There was about an eight-month lead time for boats to be built. We couldn't foresee that the crabs would vanish before the boats were built." Alaska commercial fishermen chronically complain that most of the profits from fishing go to the processors. So CFAB financed a fish processing consortium started by fishermen. But two thirds of the fishermen who were members of the limited partnership refused to sell to their own firm as originally intended. They wouldn't accept credit and sold their salmon for cash to a competing processor instead. That put their company under, taking down the state's $3 million investment with it.

Other large ill-fated loans were made to timber companies, including the Haida Native corporation. Neither the lender nor the borrower read the fading timber market correctly, and the Haida corporation was thrust into bankruptcy, owing CFAB millions.

Naive judgments were probably even more costly than market factors. For example, agency officers concede they were "needlessly taken" for huge sums by an outfit known as the Owens Drilling Company, run by a Kentucky horse farm owner named Tom Owens.

Owens impressed CFAB officers with his plan to build roads and bridges for loggers eager to cut the prime timber in southeast Alaska. The vision of a year-round industry with stable jobs was so compelling that they continued to loan him money without taking even basic safeguards to protect the loans. "Owens would get a contract for, say, eight hundred thousand dollars to do roadwork for a logging company and we would give him the money on the spot," Ed Crane, the current CFAB chief executive, recalls. "Normally, you would give him a small sum up front and then pay the rest in stages as he completes the work. But he would get it all, buy some equipment, and go to a bank and get another loan on the equipment he bought with our money." CFAB was left holding the bag for millions in unpaid loans when Owens filed for bankruptcy. The cooperative investigated whether some of the Alaska money might have been misapplied to his horse farm. But it found he had insulated his Kentucky property as well as many of his other assets.

By the end of 1984, CFAB was tottering on the verge of bankruptcy. The Spokane bank sent up Crane, then a West Coast bank consultant, to investigate. He stayed on as the new president with orders to cut everything he could to save the agency. While Crane restored a sense of stability, he could not allay Senator Sturgulewski's worst fears. CFAB announced in 1985 that even with severe retrenchment it saw no way it could live up to its agreement to repay the $32 million in interest-free seed money advanced by the state of Alaska.

Part of the prevailing wisdom was that if Alaska was to become self-sufficient, it must look to the future. Consultants, engineers, and contractors converged on the state upon hearing of Alaska's spending spree. They convinced powerful legislators that the future demanded the creation of an alternative supply of energy to feed the state's coming industrial engines when oil ran out and the cost of other traditional fuels became prohibitive.

Such was the public explanation for the plan to build a hydroelectric project at Tyee Lake, in the sparsely populated alpine area near the head of the Bradfield Canal southeast of Wrangell. This was to be the state's entry into developing water power, a model for others in this land of abundant lakes and rivers. The first beneficiaries would be the small fishing and lumbering towns on the thin panhandle of southeastern Alaska that buffers Canada from the Pacific.

In 1979, the newly constituted Alaska Power Authority submitted a license application for the project to the federal Energy Regulatory Commission, estimating a cost of $39.6 million. Consultants recommended design changes the next year, raising the price tag to $51 million. The following year the first bids came in 90 percent higher than expected, inflating estimates to $96.7 million. Governor Hammond asked no questions about the alarming cost escalation. Having accepted his share of the oil windfall to spend as his administration pleased, he had abdicated control of the rest of the budget.

The legislature was silent for a different reason. Its leaders had made a deal to fund Tyee Lake in exchange for the southeast legislative delegation's support of a much more massive hydroelectric project in the works for Susitna, north of Anchorage. Ed Dankworth, a powerful senate leader from Anchorage, had even imperiously authored what legislators referred to as the "Susitna blackmail clause." It made appropriations for a four-dam pool in the southeast, of which Tyee Lake was the centerpiece, contingent on support for Susitna.

Only Ron Lehr, a banker and a new member of the power authority, spoke up: Had Tyee Lake ever been subjected to the Division of Budget Management review requirements? It hadn't. Had the state explored the cost of alternative power sources in view of the rising estimates of Tyee Lake? It hadn't. Had the state negotiated agreements to sell Tyee Lake power to even one utility beforehand?

It hadn't. "I saw it didn't make sense to build it, but I wasn't able to stop it," Lehr says. "There were trade-offs in all these things."

The final cost of the southeast four-dam project was $450 million. But the local utilities had use for only 30 percent of the dam's over-built generating capacity. As the only customers, they had the state power authority over a barrel. They were able to negotiate subsidized rates so low that the state had no hope of ever recovering its investment equity. Fortunately, the massive Susitna dam, the cost of which had been projected at $15 billion, was never built. As oil prices started dropping, so did enthusiasm for the plan. But the decision cost the state $350 million in consulting fees and engineering studies.

The Alaskan public as well as its politicians deserves blame for the political exploitation and abject waste of those billions of dollars of Alaska's oil wealth. Had it been tax dollars that were squandered, the public outcry would have been deafening. But why fret over abuse of Big Oil's bankroll? Furthermore, people were too occupied with getting their own share of those dollars to complain.

State representative Ramona Barnes, a savvy, chain-smoking, sometimes raucous blonde, recalls the swell of voices across the state demanding that the oil windfall benefit all Alaskans, not just the special interests. Memories were still fresh of the $900 million jackpot from the Prudhoe Bay oil leases in 1969. Although the money did finance many worthy projects, the public perception was that it had been squandered. And people were aware that there was no master plan in place to deal with the new bonuses to the state when Prudhoe began to produce.

Barnes says she was among the legislators who demanded an end to the state income tax. But Governor Hammond, his expensive agricultural program just starting, wanted to keep the income tax and instead invest some of the oil money in a permanent fund that would pay future dividends to Alaskans. The money-flush legislature accommodated both. It repealed the income tax in 1981 and created a permanent fund in 1982. A voter referendum in 1976 had overwhelmingly amended the constitution to create a savings entity. Now the legislature gave it form by assigning 25 percent of the revenues from the mineral leases and oil royalties to a gigantic public trust. Further, it specifically protected the principal from future administrations or legislatures that might be inclined to raid it.

In 1982, the state sent a $1,000 check to every man, woman, and child deemed a citizen of Alaska—the first distribution of Alaska's Permanent Fund. (Among the recipients were untold numbers of Outsiders who had established mailing addresses in the state at the first whiff of the dividend program.) Since that year, Alaska has taken in no taxes from individuals. It has no income tax, no sales tax, and no wage tax. (Some Alaskan cities, however, have enacted their own sales taxes.) Instead, the state distributes a yearly Permanent Fund payment to all who have lived there for at least the previous two years. Though some public officials have tried to circumvent its sanctity over the years, the fund remains untouchable and its assets grow steadily each year. The dividends, which amounted to $952.63 a person in 1990, are something of a mandated relief program for rich and poor alike, unparalleled in the world. The state has also added an unprecedented $250-a-month bonus program for every citizen sixty-five years of age or over.

The state's population increased rapidly as people got news of its good fortune and generosity, from 414,000 people when oil started flowing in 1979 to 542,000 by 1986. In Anchorage, which grew more than 40 percent in that time, housing prices soared and living space became so scarce that even shacks and lean-tos became rentable property. The clamor for state-funded housing assistance became loud in the land.

William Parker, a self-described "early hippie" who served as a legislator in the early eighties, recalls the lament that young couples could not afford to live in Alaska. How could they be expected to stay and build the new state if they couldn't buy a house? "It was true," Parker says. "The smallest houses were going for a hundred thousand dollars in Anchorage and double-digit interest rates were driving mortgages out of sight. The banks wouldn't lend unless you came in with a good credit record and a big down payment, which most people here didn't have."

Alaska's legislature sympathized. For starters, it voted a $560 million appropriation in 1980 to subsidize mortgage interest rates, reduce down payments, and extend the time to pay. Fueled with the state's oil money, the Alaska Housing Finance Corporation (AHFC) moved in where no private lending institution would tread. Banks and the other traditional mortgage lenders in the early eighties were asking 15 percent interest, 20 percent down, twenty years to pay. The amenable AHFC enabled Alaskans to buy a house with a 10 percent mortgage, 5 percent down, and twenty-five years to pay.

Realtors, builders, and financial institutions made fortunes in the real estate boom that followed. The state's fifteen banks did particularly well. They received the applications, processed the loans, and sent them to the AHFC for virtually automatic approval. In return they received a generous 1 percent service fee and none of the risk—the state guaranteed the mortgages. The AHFC bought almost all the new mortgages written in the state during the first half of the 1980s, underwriting more than $1.3 billion in housing loans in 1984 alone.

Jack Linton, an early-vintage Alaskan with a long career in banking, was one of the first executive directors of the agency. He remembers the political meddling that accompanied the free flow of state money into the housing market. "Since the state legislature was providing the funding, or the backup loans, the politicians wanted a finger in the pot," Linton says. "You would get a call from a senator who was complaining that his brother-in-law was turned down for a loan. I was supposed to look into it and make sure he got it." When he balked, Linton says, life soon became intolerable.

The breaking point came when politicians tried to force the agency to bend its liberal lending policies to take care of housing for one more segment of the population, those with the lowest incomes. Many of them were decent people struggling to get a start in life, but others were financial deadbeats and recent arrivals with credit records so vague they couldn't qualify for a car loan. But they were helping build the new state, too, the argument went. Didn't they also deserve to own homes?

The mobile-home dealers, a powerful special-interest group in Alaska, thought so. They pressed a plan upon their friendly legislators: Finance mobile homes with the same generous terms as real estate. Never mind that banks treat mobile homes as personal property with a far shorter life than real estate. Or that they demand far stiffer terms—25 percent down and seven years to pay. The enabling legislation gave the AHFC the flexibility to set its own financing rules, and the politicians knew it.

Linton says that the cochairman of the house finance committee, Russell Meekins, Jr., called him to Juneau and told him that if the AHFC was interested in its appropriation, it had better accept the proposed terms for mobile homes. He was also approached by the lieutenant governor, Terry Miller, and John Sackett, the chairman of finance in the senate, with the same request, he adds.

Linton says he told them he wouldn't do it because it violated all prudent lending practices. "Next, the chairman of the AHFC board calls. 'We have to do it,' he insists. 'If we want our appropriation, we have to do it.' They did it. Whereupon I resigned."

The surge in mobile-home buying in Alaska created a fortune for dealers, benefiting among others the family of Russell Meekins, Jr. Meekins's father ran a mobile-home business in Anchorage.

Meekins, who abruptly quit Alaska politics after his term expired, now lives in Wellesley, Massachusetts, where he is a prominent civic leader. He defends his support of mobile-home financing and denies that he ever threatened to hold up AHFC appropriations or that his father's business affairs had any influence on his legislative decisions: "My actions were philosophically consistent with my record to help the little guy whenever I could."

All in all, Alaska financed a total of $6.1 billion in housing loans before the boom ran out of steam.

In 1986, world oil prices plunged from about $27 a barrel to less than $10. With the collapse came a deluge of loan defaults and bank failures. Only six of Alaska's fifteen banks survived. The boom parasites loaded up their pickup trucks and headed down the Alaska Highway to the lower forty-eight, leaving their houses with clothes in the closets, food in the refrigerators, and, often, pets in the backyards.

The state lost $1.128 billion in mortgage foreclosures by the end of the 1980s. Not surprisingly, the highest rate of defaults occurred among mobile-home owners. The financial damage to the state was relatively small in the context of the total losses that it had to absorb, but the battle that Linton had lost cost the state dearly in another way. Randy Boyd, who handled hundreds of trailer foreclosures for Alaska State Bank, remembers, "They would sell the washer-dryer, the kitchen chairs and table—all of which came with original purchase. That was their money out of town. Then some flooded the place before they left, smashed holes through the flimsy walls, or just trashed everything." A blight of deteriorating, abandoned trailers still mars the Alaska landscape, a jolt to visitors who come expecting pristine wilderness. It is a form of oil spill, state induced, a reminder of a foolhardy era.

The shining survivor is the Alaska Permanent Fund. It has assets of nearly $13 billion today and has grown each year, largely because it places its money in safe securities outside the state. It is prohibited by law from investing more than a quarter of its assets in Alaskan industries. Spared the obligation of helping Alaskan enterprises, the fund has built an impressive portfolio of bonds, government securities, and blue-chip stocks. Much of the credit for the shrewd stewardship of the fund goes to Dave Rose, the first executive director, who left his position in the spring of 1992 to start his own financial consulting firm.

Alaska is the only state in the union that pays instead of taxes its citizens. Some fear this kind of coddling may hurt Alaska in time because it keeps the shiftless and greedy in the state. But for now there is nothing rarer in American state government.

CHAPTER 10
Corruption, Alaska Size

A BRIGHT GLOW from the Prudhoe Bay oil pumping stations light up much of Alaska's North Slope during the winter darkness, but it runs out well before you get to Barrow. That's when ice fog generally takes over, as it does on this day late in November 1990, while our MarkAir pilot hunts for the Barrow airport. We land through the gloom just after noontime, and it is no lighter on the ground. I can see only a blur of a town. Its buildings are dim outlines in clouds of swirling snow and frozen steam from the exhaust of trucks and cars that are kept running even when people stop for lunch, for fear it will be impossible to restart them.

It is cold in Barrow in November. The temperature is twenty-nine below with a windchill factor of sixty below. As the northernmost city in Alaska, Barrow sits on the edge of the Arctic Ocean, hemmed in by ice much of the year. The sun disappears in mid-November and doesn't appear again until mid-January. Polar bears sometimes come into town across the packed ice floes in search of defenseless dogs tied in their pens.

It costs $915 round-trip to fly to Anchorage from here, bootleggers charge $50 for a fifth of whiskey, and gasoline is $2.80 a gallon. You can't drive to Barrow. And nobody drives very far around Barrow—the longest road goes only 12 miles. "Why would anyone want to live here?" I wonder. Yet thirty-three hundred people do.

"It's peaceful here," replies Marie Adams, an amiable, college-educated Inupiat whose father was a reindeer herder. One of twelve children, Marie got through Evangel College in Missouri with financial help from the Assembly of God church, but she chose to return to her subsistence family in Barrow instead of staying with a good job in the lower forty-eight. She is a dedicated whale researcher and recently opened Barrow's first public information agency, which publishes a newsletter on community activities.

Marie is proud that her people have survived for thousands of years in what some say is the world's most hostile inhabited environment. Each generation handed down the skills to capture migrating caribou, net returning salmon, and identify edible plants on the tundra, among other necessities for survival. The pride of the Inupiat to this day is the harpooning of the giant bowhead whale. In 1977, the International Whaling Commission tried to put a moratorium on the taking of bowhead whales, but so intense was the Eskimo protest that U.S. authorities helped the North Slope borough obtain special quotas for subsistence harvest. Each catch of these monsters (they weigh as much as fifty tons) is shared among the villagers and is an occasion for festivals and prayers.

The Eskimos in Barrow escaped contact with the Russian trappers who came in the eighteenth century. Other Alaska Natives were not so lucky. The invaders decimated the sea otter population and often enslaved the aboriginals who lived in the Aleutians. But the Inupiats had to overcome the hardships inflicted on their food chain by flotillas of American whaling ships that all but eliminated the whales that migrate in the Arctic. Yet they continue to hang on to their harsh world, to the perplexity of anthropologists and historians.

A visitor to Barrow quickly senses that Marie and much of the Native community are fighting harder than ever to save their culture. Over the past fifteen years, the threat to their traditional way of life has been greater than any they endured through centuries of famine, disease, and natural disaster. The discovery of oil on the North Slope has transformed these once primitive people into privileged people. Their land has created billions of dollars of wealth, and they have had serious problems dealing with it.

As the seat of the North Slope borough, which taxes all of the Prudhoe Bay oil fields, Barrow is the richest city, per capita, in the United States, and possibly in the world. In recent years it has also probably attracted the greatest number of unscrupulous people, per capita. That they have managed to extort many millions of dollars of Eskimo wealth is a scandal little known beyond Alaska.

Oil companies did their utmost to prevent the incorporation of the North Slope borough. Early in the 1970s, they sought injunctions in court, arguing that it did not make sense to create a municipality of 56.6 million acres (about the size of the state of Minnesota) just so a small band of once nomadic people could tax the Prudhoe Bay oil complex, plus every future oil well, the pipeline, and service centers across the entire top of Alaska. "Who Will Control the Dazzling North Slope Wealth?" one Anchorage newspaper headlined the battle.

With the several thousand oil company workers who had recently arrived in Alaska barred from voting, an essentially Eskimo electorate voted 402–27 in June 1972 to create the North Slope borough. Citing the terms of the state constitution, the Alaska courts validated the election, thereby certifying the largest local government in the world. The borough starts where Alaska meets the Chukchi Sea on the west and encompasses the entire oil-rich tundra north of the Brooks Range to the Canadian border on the east.

The fifty-seven hundred predominantly Inupiat Eskimos who occupied the eight isolated villages in this desolate land probably had no idea of the wealth that would be theirs. Newspapers estimated that the oil company facilities at Prudhoe Bay alone would swell the assessed value of taxable property to $10 billion. It reached $13.6 billion by 1987. Meanwhile, Outsiders scrambled even before the vote was in to advise the Eskimos on how such a solid tax base could be used to create instant millions for themselves, and, of course, their advisers. Investment houses were begging to sell municipal bonds backed with such guarantees of oil revenues.

The borough elected Eben Hopson, an Inupiat whaling captain and respected patriarch, as its first mayor. He had a clear vision on how to apply the money. "The caribou and the whale have formed the base of our existence, but cash has become a way of life," he stated. "We have been introduced to dwellings heated by oil, with running water and even indoor toilets." He pledged that the people of the North Slope henceforth would have all the amenities of Anchorage or Fairbanks. No longer would they have to carry their wastes to village tank trucks or hunt for blocks of blue ice to melt for drinking water. And never again would they have to worry about the caribou not returning or the salmon disappearing. There would be well-paying cash jobs for everyone.

As the center of his improvement program, Hopson authorized the Barrow Utilidor, an extraordinary sewer and water system that would be installed in heated tunnels burrowed beneath the permafrost. He promised to connect it to every home in the city. To give his people the finest education, Hopson started building the largest and most up-to-date high school in Alaska. It would occupy 119,532 square feet, and include an Olympic-size swimming pool, a college-size basketball auditorium, plus separate rooms for wrestling, gymnastics, and karate—an impressive facility for a district that averages about 250 high school students.

Hopson died in 1980. He did not live long enough to see either project completed. Nor did he live to learn of the massive corruption his well-intentioned building program would spawn. That sad distinction belongs to the administration of Mayor Eugene Brower, a boyish Inupiat who took office in 1981 at the age of thirty-three. Brower, also a whaling captain, had been the public works director, but had grown impatient with the slow pace of the capital improvements started by Hopson and the interim mayor, Jake Adams, another whaling captain. (Barrow has 144 whaling captains.) When the baton passed to Brower, he set out to take Barrow from the Stone Age to the space age as fast as he could get vessels and planes to ship the lumber and hardware from Seattle.

Coached by outsiders, Brower borrowed hundreds of millions of dollars—the debt swelled from $453 million to $1.2 billion during the three years he served—by floating bonds backed by the borough's oil property-tax base. This was quite a spectacular feat for a municipality of fewer than six thousand people. Moody's Investors Service rated the bond issues A or better. Wall Street bond brokers and their Alaskan agents feasted.

As the debt began to approach that of Philadelphia and other large cities, several alarmed lawmakers started to call for legislation to put a cap on Barrow's runaway borrowing. They feared the state could be left paying off the debt in the long term when oil revenues started declining and Barrow could no longer meet its payments. In response to these threats from Juneau, Brower signed multimillion-dollar construction contracts the way some mayors sign proclamations. And with about the same amount of scrutiny.

Brower, whose education was limited to BIA schools, could not have masterminded so massive a capital program by himself. Experts of all stripes from the Outside streamed in to help, and Natives often stopped to admire their shiny Lear Jets and Cessnas parked at the airport. But there were two figures who emerged from the shadows of Alaska's political life to grab the inside track from the day that Brower took office.

Lew Dischner, then sixty-five, was a portly, glad-handing political veteran who had been part of the power structure in Juneau for as long as Alaska had been a state. He had been appointed the state's first commissioner of labor in 1959 by Governor William Egan, a Democrat. Dischner left the post after a year and resurfaced in Juneau as a lobbyist for several large clients, including the Teamsters, the most powerful labor group in the state. Along the way, Dischner demonstrated an adeptness at getting government loans and state contracts to build a private business empire. He developed a waterfront mall in Juneau, won the contract to operate a laundromat in a state-owned residential high rise, and acquired a hotel that leased rooms to the state.

Dischner also built a network of influence. With the help of the Teamsters, he became a leading fund-raiser for the Democratic party and certain helpful Republican candidates. His ability to deliver hefty campaign contributions earned him a reservoir of return favors at all levels of local and state government, to be tapped when needed.

With the birth of the North Slope borough in 1972, Dischner headed north to the future. He arrived bearing a campaign chest for Eben Hopson, who rewarded Dischner by hiring him as a borough lobbyist. When Hopson died, borough assembly president Adams served out his term and was expected to win the next election, in October

1981. But Dischner decided the next mayor should be Brower, the young Eskimo rising in the public works department.

Dischner masterminded Brower's campaign, raising $100,000 to fund it. Much of the money was laundered, given in the names of a variety of people but actually coming from Dischner and an assortment of slope contractors who expected to be rewarded once their man was safely in office.

It was a daring gamble. Brower won by only twenty-four votes, but he lost no time in rewarding Dischner. He made the lobbyist consultant to the mayor on capital improvement projects, a lucrative combination that would set Dischner up to make $250,000 a year for lobbying services alone, two and a half times Brower's own salary.

The mayor's other confidant was Carl Mathisen, then forty-nine, a small-time political wheeler-dealer who dripped gold. As glum as Dischner was cheerful, Mathisen made his statement with what must have been a record-size wristwatch studded with gold nuggets, featuring two twenty-karat bears growling at each other across the timepiece. Mathisen had worked in the 1970s for the borough's bond counsel, Bob Dupre of Juneau. While helping Dupre reap a bounty of commissions by floating the $1.2 billion in bond issues, Mathisen had been laying the foundation of his own future. Though short on formal education, Mathisen had enough contracting experience in Anchorage to understand the ways of government and bureaucrats. He convinced Mayor Hopson to hire him as borough training program coordinator. That was how he got to know Brower in the public works department.

Mathisen became Brower's mentor, teaching the young Inupiat ways of government not found in textbooks. Mathisen also ingratiated himself with the Brower family, showering them with gifts and even became godfather to a Brower son. Brower reciprocated by persuading his father, yet another whaling captain, to appoint Mathisen a member of his renowned crew. The only white man ever to be so honored, Mathisen bowed out after he had a terrifying nightmare about a polar bear in camp on his first outing and injured his leg on the second.

But when Brower ran for mayor in 1981, Mathisen and his wife were in the forefront of his supporters, with a $5,000 check. One of

Brower's first acts in office was to install Mathisen as a consultant to the mayor and to the public works director handling capital projects, a post that would pay him an average of $300,000 a year. His initial consulting contract paid him $156,500.

The two consultants were well paired. Dischner was skilled in siphoning money and Mathisen had the nuts-and-bolts savvy to provide respectable projects. Together they recruited a compatible cast of engineers, architects, technicians, and construction and service companies to do their business. Many were struggling specialty firms and some did not exist until the borough started letting contracts. But all were expected to abide by the house rules, which required them to systematically kick back 10 percent of every borough contract to Dischner and Mathisen.

Awed by the way his two advisers got things moving, Brower didn't question them when they picked a Seattle firm, the H.W. Blackstock Company, to purchase all materials, all shipping, and anything else that had to do with servicing the borough's capital construction projects. As an influence peddler, Dischner had had a long relationship with Blackstock. He took good care of his client.

Dischner engineered a contract that permitted Blackstock to tack 30 percent on top of the cost of anything it provided, an arrangement unheard of in municipal government or private industry. The president of the lucky company was Kenneth Rogstad, the former Republican chairman of King County, which includes Seattle. Dischner knew that Rogstad could be counted on to contribute to campaigns of well-positioned Democrats in Alaska as well.

"The consultants became the government," Chris Mello, then contract reviewer for the borough, says. A California native, barely thirty years old and fresh out of California Western School of Law in San Diego, Mello was working at his first real job. He admits he was puzzled by what he saw at first, and then was simply dismayed. "In my first meeting with Lew Dischner, he told me he was a blood brother to the mayor, and what he said went," Mello says. "Suddenly the borough was starting hundreds of projects and running them was wrested away from the borough employees and turned over to the consultants. We were reduced to clerks."

Brower signed contracts for more schools, firehouses, health clinics, roads, worker camps, incinerators, engineering studies, and architectural renderings, plus expensive change orders for the Utilidor and other projects in progress. At one point, the sparsely populated borough's capital budget soared to $300 million a year, which rivaled what the city of Chicago then spent for the capital needs of its millions of residents. "Consultants identified the projects," Mello says. "We were spending as much as a million a day. Projects were coming so fast that those that should have been bid were not bid. I was supposed to review the contracts before they went to the mayor. But sometimes those documents would arrive in the legal department already signed by the mayor. What was the point of reviewing them then?

"We hear rumors of kickbacks. Then we find out later that Lew and Carl formed their own companies to work on projects. The next thing you know they are negotiating contracts on behalf of the borough with their own companies, and the costs of the projects go way out of line."

Harold Curran, another young law-school graduate, was also among those who felt uneasy. He had come to Alaska as a VISTA volunteer and followed his girlfriend, an environmental planner with the Trustees for Alaska, a nonprofit public interest law firm specializing in Alaska environmental issues, to a project in Barrow. Hopson welcomed the talent and hired him as borough attorney. "When I found out that a sole-source contract was awarded for six health clinics, I advised Brower that it should have been a bid contract," Curran says. "At one point I wrote him a memo noting that some of the contracts that had not been bid seemed to be in the high dollars. Soon after, I got a letter from Brower's litigation counsel telling me that the price of a contract was not a legal issue."

Frustrated, Curran tried to interest John Larson, news director for Channel 2 in Anchorage, Alaska's largest television station, in looking at the way the borough handed out contracts. Larson said the station didn't have the staff to investigate a situation so far away. But the Eskimos in Barrow did not need media exposure to see that things were getting out of hand at borough hall, or that Outsiders were helping themselves to millions of dollars of their money.

The Utilidor was initially estimated to cost $80 million, but Brower's change orders helped send the total up to $250 million by 1984, when only 10 percent of the buildings were hooked up. (The cost had reached $330 million by 1991, when still only slightly more than half of the homes were being serviced. It is by far the most expensive public works project ever attempted in Alaska.) The bill for the high school, initially projected at $25 million, soared to $80 million, which amounts to $320,000 per student. In his book *Alaska,* James Michener gives the impression that it is an ugly building. To me, the appearance did not register as forcefully as did the distortion of educational priorities. An expensive, life-size bronze statue of Hopson greets a visitor at the entrance, and a $75,000 mural decorates an inside wall. Yet the library is no larger than the gym's smallest exercise room, its book supply is glaringly meager, and its current periodical shelf does not contain any out-of-state newspapers or even one from the state capital.

What attracted even more public attention than the skyrocketing cost of these projects, however, was the change in the mayor's lifestyle. Brower now traveled by private jet to Anchorage, Seattle, Palm Springs, and, it was rumored, the casinos in Las Vegas. After one trip, he returned sporting a huge diamond ring and expensive new suits. The mayor's new boat, worth at least $35,000 according to local mariners, was the envy of all who boated upriver to the caribou grounds. Brower also started showing off a new Browning semiautomatic shotgun.

The prevailing wisdom is that an incumbent mayor who provides full employment and more than the usual amenities for his constituents is bound to be reelected. Brower did all that and had a reelection war chest of $250,000 to boot, raised with Dischner's expertise. But the people of the North Slope could feel their fortune being drained by Outsiders, and quiet resentment was mounting. In the fall of 1984, they voted Brower out of office. The victory of George Ahmaogak, a borough worker who is only part Inupiat (but a whaling captain, of course), was the political shocker of the year.

During Brower's last five days in office, his administration pushed through more than $15 million in checks and signed $7.6 million in contracts. The frenzy was described in a story filed from Barrow by Bill White, business editor of the *Anchorage Daily News:* " 'There are planes leaving in a half-hour. Get the checks out,' outgoing mayor Brower barked at an accounting officer. While the planes waited, the men they had brought hovered over the mayor and the accounting clerks. They wanted checks for millions of dollars and they wanted them cashed before Brower left office." Among the principal beneficiaries were Dischner, Mathisen, and Kenneth Rogstad, the head of the H. W. Blackstock Company.

Incoming mayor George Ahmaogak immediately noticed that many borough files were missing. He assumed they were shredded, whereupon he went to the assembly and asked for an audit of borough finances. The audit, performed by the Fairbanks accounting firm of Main Hurdman, surprised even those who had expected the worst. Besides finding many contracts that had not been legally bid, the report showed that millions of dollars had been spent on services for which no invoices existed. It also identified an array of contracts with "an improper scope and fee relationship," meaning that the amount paid appeared to be too excessive for the service. The auditors particularly questioned the propriety of nearly $20 million of expenditures to H. W. Blackstock. Not only were many of the payments in cash, contrary to purchasing procedures, but sometimes goods were never delivered. Among the many discrepancies cited was a $365,134 payment for borough housing development furniture. Neither invoice nor furniture appeared to have been received.

The most eye-opening finding was that Dischner and Mathisen, while on the borough payroll as consultants, had set up firms of their own to get borough business. The largest, North Slope Constructors, had been favored with many millions of dollars in contracts. It was incorporated, with Dischner, Mathisen, and Rogstad as equal owners, one year after Brower took office. North Slope Constructors was able to shut out other firms by bidding low and then negotiating change orders. "That substantially increased the size of the contract without substantially increasing work to be done," the auditors reported.

The whiff of wholesale fraud inspired newspapers in Fairbanks and Anchorage to dig deeper. They soon exposed layers of lesser players—lawyers, architects, engineers, consultants, public-works employees, and even a state senator—who had benefited from the questionable payments, kickbacks, and tainted gifts that flowed like water in the Barrow building boom. Public attention inevitably shifted to Governor William Sheffield. What was he going to do about the North Slope corruption?

The governor's first response was that he was not aware that any state money had been involved. The *Anchorage Daily News* quickly challenged him, showing that more than $4 million of state money had been budgeted for a half-dozen projects questioned in the audit.

Then Anchorage representative Fritz Pettyjohn charged that state attorney general Norman Gorsuch was dragging his feet in investigating the matter because several of those who landed lucrative contracts in Barrow were heavy political contributors to Sheffield's campaign. Dischner himself was one of Sheffield's largest contributors. A secretary in the governor's office says Dischner walked into the office one day and tried to drop off a $60,000 check for the governor's campaign committee. The secretary, recognizing the illegality of the gift, refused to accept it. But the surprise was that Rogstad, a Republican power in the Seattle area, figured so prominently in raising money for Sheffield, a Democrat. According to newspaper reports, which were not denied, employees of Blackstock sent $7,000 in checks to Sheffield and the company helped organize a fund-raiser in Seattle to reduce the sizable campaign debt Sheffield carried into office.

Suddenly, the question of whether Sheffield would act became academic. By the spring of 1985, the governor himself was in deep trouble. Someone had tipped off the *Fairbanks News-Miner* that there was something shady about a ten-year, $9.1 million lease the state had signed for office space in a Fairbanks building. The specifications had been written so narrowly that only the one building had qualified.

Stan Jones, a reporter on the *News-Miner*, dropped the Barrow story and plunged into a round-the-clock investigation of the Fairbanks lease. He reported that a labor leader named Lennie Arsenault, who had helped raise $92,000 for Sheffield's campaign, had a financial interest in the favored building. Further, he found that Arsenault had had discussions with the governor regarding the lease and that employees within the state leasing office had protested the circumvention of leasing procedures.

The stories troubled state prosecutor Dan Hickey, a Georgetown Law School graduate who, like Curran and Mello in Barrow, was among the young lawyers who had answered ads for jobs in the emerging state of Alaska. He became chief prosecutor in 1975. In spite of Sheffield's public declarations of propriety, Hickey quietly started to investigate the Barrow mess without the governor's knowledge. But other disturbing events soon demanded his full attention. Two employees from the administration's procurement office visited Hickey and bared their concerns about the Fairbanks lease. They also told him Stan Jones had filed a request under Alaska's Freedom of Information Act to see the file, which, they said, was damaging to the governor.

Crusades against corruption in government have a poor record in Alaska. In fact, the system seemed to find ways to abort aggressive law enforcement, a fact no one knew better than Hickey. In 1982, he had led a grand jury investigation into the dealings of state senator Ed Dankworth, former head of the Alaska state troopers. While cochairman of the senate finance committee, Dankworth and a business partner had bought the former Isabel Pass pipeline camp for $1 million and tried to sell it to the state for use as a correctional facility—at an asking price of $3 million. The grand jury indicted Dankworth for conflict of interest, but the courts gutted the case, ruling that the senator was protected by legislative immunity. None of his conversations or actions to persuade the legislators to appropriate money for the camp purchase could be used against him; they were considered part of the legislative process. Hickey had no choice but to dismiss the case.

Hickey was also well aware of what happened to another whistle-blower, Representative Russ Meekins, Jr., the upstart who had captured the speaker's chair in the famed 1981 legislative coup. In the spending frenzy of 1982, Meekins publicly accused state senator George Hohman (D-Bethel) of trying to bribe him with promises of a bag of money in exchange for support of a water-throwing plane. Hohman was convicted and expelled from the senate. However, ten years later he still had not paid his $10,000 fine and was gainfully employed as acting city manager of Bethel. Meanwhile, Meekins's brilliant political career suddenly came to a screeching halt. He did not run for reelection and left the state.

These were parables Hickey could not ignore as he began to probe the affairs of the highest officer in the government of Alaska. Nevertheless, he impounded all the leasing records in the procurement office and launched a grand-jury investigation.

Governor Sheffield made two appearances before the grand jury. Soon after the first, Hickey went to his boss, Attorney General Gorsuch, with an unappealing prognosis. The ramifications of the case required outside expertise. A pained Gorsuch agreed. They looked to Washington, D.C., and brought in George Frampton, who had worked as a special prosecutor with the Watergate grand jury that led to the downfall of President Nixon in 1974. Tensions in the state escalated as politicians and the public soon suspected that the inquiry into Sheffield's administration involved more than the lease of a building. The grand jury worked for ten weeks, calling in more than forty-four witnesses and preparing 161 exhibits as it built a case against Sheffield.

Meanwhile, two key state housing administrators died under circumstances many continue to find troubling. In February 1985, Bruce Moore, a department of administration services employee who complained superiors were pressuring him to process the Fairbanks building contract, was found dead on a sidewalk beneath the balcony of his condominium in Hawaii. He was an apparent suicide, though he left no note and friends noticed no indication of depression or ill health. Three months later, Lisa Rudd, a Sheffield cabinet officer who also had expressed skepticism at the building contract, died within hours of contracting a virulent bacterial infection. No evidence of foul play was found in either of the deaths, but newsman Jones and state housing employees of that era still shake their heads over the deaths and call them "an incredible coincidence."

On July 2, 1985, the grand jury returned a devastating report. It charged "a serious abuse of office" by Governor Sheffield and his chief of staff, John Shively, in their alleged intervention into the lease process and in their attempt to frustrate official investigations into the matter. Inspired, according to some, by the climate of the Watergate hearings, the jurors called Sheffield unfit to hold office and recommended the senate be called into special session to consider impeaching the governor.

Shaken by the finding, Sheffield stuck by his testimony that he did not remember meeting with Arsenault and that consolidating state offices in the Fairbanks building would have saved the state money. Nevertheless, about six hours after the grand-jury report, Gorsuch issued a legal opinion stating that the administration should cancel the lease because it was tainted by favoritism.

The attorney general, who had already said he wanted to resign, left office the next week. Sheffield quickly appointed Hal Brown, a lawyer from Ketchikan, to succeed him. One of Brown's first acts was to call in State Prosecutor Hickey. "I was told to get out quietly," Hickey says. He left, but not quietly. He issued a blistering parting message citing the work to be done, including the state's responsibility to investigate the situation at Barrow.

The senate impeachment debate took on even more strongly aspects of Watergate. The Republicans, who had 11–9 control of the Senate, hired former Watergate committee counsel Sam Dash to oversee the proceedings. Meanwhile, Sheffield also reached down to Washington for another Watergate pro, Philip Lacovara, counsel to Watergate prosecutor Leon Jaworski, to defend him.

Dash advised the senate that Sheffield's tampering with state leasing procedures might not amount to an impeachable offense, but perjury would. (The broad pattern of lying to cover up the crime rather than actual complicity in the Watergate break-in was the biggest factor in Nixon's ouster from the presidency.) Dash pointed out that Sheffield had testified four times that he could not recall ever meeting with Arsenault, while not only Arsenault but John Shively, the governor's own chief of staff, told the grand jury in detail about a meeting in which the governor and Arsenault discussed lease specifications.

Many thought Dash had made a case for impeachment, but at the moment of truth, the majority crumpled. Observers speculated that some didn't understand what was wrong with helping a campaign contributor, and others were bound to the unwritten political code that you don't stone anyone in your own circle today lest you be stoned tomorrow. Still others clearly saw official misconduct but simply did not have the courage to make a stand.

In the end, the senators not only rejected Dash's recommendation, but gutted a subsequent rules committee resolution denouncing Sheffield for questionable veracity and "significant irregularities." Instead, they called for a study into state procurement procedures and then added a rebuke to the investigators in the form of a resolution asking that the Alaska Judicial Council "study the use of the power of the grand jury to investigate and make recommendations . . . to prevent abuse and assure basic fairness." Five years later, as memories faded, the legislature even reimbursed the governor for his legal expenses. A catchall bill, ostensibly passed to fund the state's longevity bonuses and legal expenses to recover disputed royalty payments from oil companies, included a payment of $302,653 for Sheffield. And when the Democrats regained the U.S. presidency in 1992, Sheffield emerged as Governor Clinton's dispenser of patronage in Alaska.

The trauma of the Sheffield controversy plus the departure of Hickey doomed any state action against government corruption in Barrow. The new state attorney general had no appetite for searching out corruption in Bush villages, much less for the turmoil of another grand-jury investigation. By now just about every element in the protection system was prepared to do nothing about Barrow. The reason seemed clear to Roger McAniff, a University of Alaska engineering professor who, as a business consultant, had analyzed the inflated contracts let during the Brower years: "Alaska simply had no experience in dealing with white-collar crime."

But while the desire for money induces corruption, it can also counter it. The millions lost to favored contractors on the North Slope had infuriated many contractors who sought and were denied the chance to compete. Several refused to accept the affront without a fight. Reports of kickbacks for contracts and of the millions Eugene Brower had dished out during his final days in office came to the attention of the FBI. The bureau dispatched agent G. Bruce Talbert to Barrow to investigate. He came back with a harrowing story. The payouts appeared to be the desperate final stages of a sophisticated system in which a tight circle of white manipulators had been making personal fortunes by milking the newly rich Eskimo municipality. Only someone knowing how to match political greed with free-enterprise chicanery could have masterminded such a scheme, Talbert reported, and Brower's consultants, Lew Dischner and Carl Mathisen, had just those skills.

Meanwhile, North Slope's escalating bond indebtedness was receiving attention around the state. Representative Fritz Pettyjohn of Anchorage had been urging fellow legislators in Juneau to support a bill to cap the borough's extraordinary borrowing spree, lest the state eventually get stuck with the debt. Some Natives assailed Pettyjohn as a racist for denying Eskimos the right to use their own money to upgrade their living conditions, but their complaint was unlikely to pacify an uneasy legislature for long.

By the spring of 1985, the FBI, now joined by Internal Revenue Service investigators, was able to piece together a picture of the dimension of corruption. The bureau handed the Justice Department a bribery, extortion, and fraud case the likes of which the nation had rarely seen before.

Alaska's U.S. attorney was a tall, trim, low-key lawyer in his late thirties named Mike Spaan. His training was solid—Boalt Hall law school at the University of California Berkeley, two years as a legal assistant for Alaska's senior U.S. senator, Republican Ted Stevens, and six years with a private law firm. But he also was very conscious of whose turf he was treading on. While many of his peers would have jumped at the chance to establish a political reputation as a foe of crime and corruption, Spaan was reluctant to preempt the state. Further, he felt state laws could be applied more directly and effectively than the Racketeer Influenced and Corrupt Organizations (RICO) Act, the federal antiracketeering law that covers local corruption.

But the FBI gave Spaan little choice. It was possible, the bureau reported, that Alaska was sitting on the biggest bribery, extortion, and tax evasion case in municipal history. The FBI identified more than a dozen people—public officials, contractors, engineers, architects, and assorted lesser players—as members of Barrow's inner circle who had enjoyed grossly inflated contracts in return for which they'd kicked back a steady stream of millions to the masterminds in the mayor's office. The bureau wanted the authority to start subpoenaing records.

Whatever the difficulties of using federal law to crack a state corruption case, Spaan was persuaded. Appalled at the amounts of money and instances of public betrayal, he was further dismayed to find he did not have the staff to move against everyone at once. He could prosecute only in stages and hope that he could finish the job before the five-year statutes of limitations ran out. While much of the state's attention was still occupied by the Sheffield impeachment battle, Spaan moved quietly to start a federal grand-jury investigation into the North Slope mess.

Early in 1985, Spaan and Talbert began mapping out the first full-scale white-collar corruption case ever prosecuted by the U.S. attorney's office in Alaska. Spaan decided to move first against the lesser figures, hoping he could get them to plea-bargain for lighter sentences in exchange for a promise to testify for the government in the bigger cases.

On May 30, 1986, the federal grand jury finally handed down its first indictment. After more than a year of news reports of intense witness traffic in the federal building, it was an anticlimax. A lone Inupiat named Irving Igtanloc, Brower's public works director, was indicted on six counts of extortion, wire fraud, and income tax evasion. Nor was the substance of the charges particularly earthshaking. Igtanloc was accused of extorting gifts from a Washington State firm doing business with the borough, to wit: a remodeling job on his home, a .44-caliber revolver, and the mounting of an 18-inch lake trout his wife had caught while they were guests of the contractor at a fishing lodge.

Quite incidentally, Spaan announced, the man charged with providing the gifts to Igtanloc had been indicted for bribery but had pleaded guilty to a reduced charge in exchange for an agreement to cooperate with investigators. The lesser charge—illegal shipment of gifts of liquor—was laughable, especially in Alaska, but Spaan felt he had at least made a crack in the inner circle.

The name of the plea bargainer was Joseph P. Brock, and he had been a $250,000-a-year executive for a consulting group called MMCW. Based in Anchorage, the firm was a partnership of McCool McDonald Architects of Seattle and Anchorage and the Bellevue, Washington, engineering firm of Coffman and White. Dubbed "the engineers" in the FBI investigation, MMCW was a major player in the borough's capital improvement program. It had been handed an annual $7.8 million, no-bid contract to administer the technical side of the project development. MMCW determined the scope of the work and the estimated cost, then placed the contracts. Many of the contracts, the FBI found, were steered to architects and engineers within the MMCW family, at suspiciously lucrative terms. Brock was an important point man for the MMCW operation. But his feeble penalty—two years' probation and a $4,000 fine—made some wonder whether the state was in for another round of typical Alaska white-collar justice.

By the end of 1986, the indictments had picked up speed, and a picture of looting in the North Slope had finally begun to emerge. Martin Farrell, Brower's top lawyer, was charged with fraud for getting Brower to sign a $720,000 contract, with a $150,000 nonrefundable advance, two days before leaving office in 1984. Thomas Gittins, a contractor, was charged with kicking back 10 percent on all borough contracts to consultant Lew Dischner and with doing a $600,000 home remodeling job—at Dischner's expense—on the home of Myron Igtanloc, Irving's brother. Myron maintained an 8,500-square-foot Anchorage residence with gold-plated plumbing fixtures while serving as the North Slope's capital improvement projects coordinator. Myron Igtanloc himself was charged with three felony counts of tax evasion for failing to report $600,000 in income over a three-year period.

It was also disclosed at this point that Gittins claimed he had been paying an additional 3 percent on all contracts to a North Slope Native legislator, Al Adams, who soon after had been elected to the state senate. He at first insisted he had received "a little more than his legislative pay"—$45,000—for helping Gittins get contracts, most of which were funded by public money. Subsequent testimony would charge that Adams received at least $700,000 for services described as "public relations, ensuring compliance with local hire, and such other duties as Gittins might assign him." Those allegations did not figure into the indictment, nor did the Alaska state legislature find any breach of ethics after it made a cursory investigation.

The North Slope grand-jury investigation went on for two years before it touched principal figures. On February 2, 1987, ex-mayor Eugene Brower was indicted on fourteen counts of receiving bribes, gifts, and loans from contractors and lobbyists in return for preferential treatment. Several big names—Dischner, Mathisen, and Rogstad among them—were identified as having provided the favors, but they were not indicted, to the consternation of the public and the press. Spaan dodged the questions. As it turned out, he was buying time to negotiate with Brower in the hope that the ex-mayor would agree to reduced charges in exchange for testifying against those who extracted the heavy money.

For several months the ex-mayor held out, vowing to prove his innocence in court. Finally FBI agent Brent Rasmussen, who had worked with Talbert on the investigation, set up a meeting with Brower and showed him the government's evidence of the "10 percent clause," the secret deal Dischner and Mathisen had with the engineers, architects, and contractors under which the latter kicked back 10 percent of their payments from the borough. "I don't know if words can describe it," Rasmussen said. "Brower went into a state of shock. He broke down entirely, he was sobbing, and the interview had to stop. He was unable to continue."

Soon after, Brower pleaded guilty to a single charge of tax evasion and agreed to help the government. This time there was not only the uneasy feeling that a truly major player in the corruption case was getting off with a sweetheart deal, but a lingering question in the public's mind whether Brower might have feigned his remorse.

Nevertheless, Spaan was convinced he now had what the case needed—an Eskimo leader to testify against the white exploiters. On November 10, 1987, Dischner and Mathisen were indicted. The government alleged that their illegal take from the borough amounted to $21 million. Each was charged with thirty-six racketeering counts of fraud, bribery, kickbacks, and extortion in the North Slope borough.

The slope inquiry had so drained the short-staffed U.S. attorney's office that the state attorney general agreed to lend Spaan two lawyers from his staff to help out. An attorney named Peter Gamache had been assigned there earlier and now a second was urgently needed.

Karen Loeffler got the call. A petite, dark-haired, cheerful young woman only four years out of Harvard Law School, she had come to Alaska when she became bored with her job at a private law firm in Minneapolis. She landed a job with the oil and gas section of the state attorney general's office in 1985, when the gubernatorial impeachment controversy dominated the news. She had heard only a little about the North Slope, and since the state had at that point decided not to pursue the case, she had never dreamed she would become part of it.

As soon as she reported for duty, Spaan handed her a pile of grand-jury transcripts, exhibits, and financial records with instructions to read them thoroughly. The tone in his voice said, "From now on, Karen, the North Slope corruption case is your life."

Loeffler had limited experience as a prosecutor and very little exposure to criminal law. But she did not have to read very deeply into the North Slope file to discover she had been handed an incredible mission. For the next year, she devoted long days, nights, and weekends trying to put together blocks of the evidence in a way a jury of ordinary citizens could understand. She evaluated the grand-jury testimony of hundreds of witnesses. She pored over reports of FBI agents, often meeting with them and Peter Gamache to analyze prospective witnesses. She buried herself in a small mountain of documents to unravel the workings of complicated interlocking corporations.

On October 14, 1988, the racketeering trial of Lewis Dischner and Carl Mathisen finally began. In a crowded courtroom on the second floor of the federal building in Anchorage, U.S. attorney Mike Spaan stationed his erect six-foot-four-inch frame before the jury. "This is a case about fraud, bribery, kickbacks, extortion, and corruption," he said. "The sticky fingers of Dischner and Mathisen managed to grab twenty-one million dollars in public funds from the borough in just three years. We are going to prove that twenty-one million dollars dollar by dollar."

Mathisen smiled weakly. Dischner leaned over to confer with his attorney, Douglas Pope, giving the impression he was puzzled by what Spaan was saying.

"This entire case," Pope countered, "is here because the prosecution misunderstands the situation on the North Slope." The government was confused about the law, he said, and about the duties of Lew Dischner, "a man with a strong sense of values being unfairly prosecuted for his strong commitment to the people of the North

Slope borough." Taking his turn, Laurence Finegold, Mathisen's attorney, declared, "This isn't a case about money. It is far more complex than the government wishes to have you believe. . . . It was not illegal payments that created expensive projects, but the [arctic] conditions. . . . In less than fifteen months Mathisen helped bring dramatic, if expensive, changes to the North Slope." The lawyers took about four hours to outline their cases, and then the court recessed for the weekend.

Brower was the government's first witness. In a hesitant, almost sheepish manner, Brower testified that while in office he had been showered with gifts by Dischner, Mathisen, and their contractor friends. Dischner provided him with the use of a home in a fashionable section of Anchorage. Brower also admitted to accepting a $45,000 diamond ring from him. Mathisen had paid for a customized 27-foot dory valued at $35,000. Brower's benefactors flew him to Las Vegas several times and on two occasions handed him $1,000 in gambling money. Contractors flew him to Palm Springs, where he was provided the use of a Cadillac, and he never asked who paid his hotel bills and airfare to Hawaii.

Brower admitted that many of the borough contracts he had approved were issued to companies totally or partly owned by Dischner and Mathisen. He conceded that he had pretended not to know of their involvement when he testified before the grand jury and had lied under oath to protect them. Yes, he said, he had known about Mathisen's ownership of Alaska Management Services, about Dischner's part ownership of North Coast Mechanical, and about both defendants' ownership shares in Igloo Leasing and North Slope Constructors, all big beneficiaries of Brower's public-works spending. Under questioning by Spaan, Brower even said he was aware that Dischner and Mathisen had started a jet service, called Tri-Leasing, with Dana Pruhs, a lobbyist for the nationally known Enserch contracting firm, which was based in Texas but also had sizable operations in Alaska. He conceded he had heard that North Slope contractors had to use the service or lose future contracts. (According to the indictment, Tri-Leasing extorted $570,979 from borough contractors in 1984.)

In cross-examination, however, the defense attorneys went to work. Incredibly, they got Brower not only to deny that the gifts were bribes or were used to gain his approval of contracts but to agree that the borough had benefited from them. By providing Brower with a house in Anchorage, where he had the free use of a telephone and a Cadillac, hadn't the defendants saved the borough money when the mayor was away on business? "Yes," Brower answered.

Lawyer Douglas Pope even sought to portray the gifts as part of an Eskimo *pamaq*, a Native gift-giving ritual that signifies a partner for life, with no expectation of something in return. Longtime Alaskans in the courtroom had to suppress chuckles. They knew that *pamaq* is based on helping those in need when food runs out or when disability restricts subsistence activity. A $45,000 diamond ring, a $35,000 dory, and $1,000 handouts to gamble in Las Vegas hardly sounded like aid to a Native in distress.

In short order, the defense attorneys reduced Brower to total confusion. Suddenly, what had seemed an open-and-shut case of bribery had been recast as big government coming down on well-meaning people who had tried to befriend the Eskimos on the North Slope.

The case became a trial of legal finesse. Spaan concentrated on building a paper trail showing calculated bribery. He placed in evidence an array of official records, painstakingly assembled by Karen Loeffler and IRS agent Ronald G. Chan, to show how money flowed to the borough, then to the contractors, and then in 10 percent increments to the bank account Dischner shared with Mathisen.

Spaan got some unexpected help from witnesses who suddenly began talking to save their own skins. Charles Hinson, a construction project engineer with Coffman and White of Seattle, which was part of MMCW, reluctantly testified that the engineering firm paid Dischner $1.3 million between 1981 and 1984 — an amount equal to 10 percent of its engineering fees to the borough. Geoffrey Fowler, another member of the inner circle of consulting engineers, gave an even more vivid account of what it was like to do business on the North Slope. Testifying with immunity after having pleaded guilty the year before to a single charge of bribery, Fowler said his original firm, Frank Moolin and Associates, was down to a single employee in 1980 when he got a call from his brother to come to the North Slope borough, which was having design troubles with the underground Utilidor. It turned out to be a timely call. Fowler's struggling firm eventually redesigned the entire $330 million project and emerged as a flourishing business, but he had to play by the rules. Fowler testified that besides paying bribes to Dischner, he bought guns, portable dishwashers, and tape players for borough officials, raised campaign funds, and laundered campaign money at Dischner's behest, fully aware he was breaking laws.

Thomas Gittins, a self-made contractor who started as a janitor, already had his own indictment for bribery to worry about, but he testified without immunity to assist the government. He said he met Dischner and Mathisen in Palm Springs and asked them about lining up work in the North Slope. During the discussion, he said, Dischner agreed to represent him—for 10 percent. Gittins already was allegedly paying legislator Al Adams 3 percent on all contracts, but, according to Spaan, he needed Dischner for additional protection. Gittins was rewarded with a $6,155,000 service-area project that included building a well-appointed barracks for oil field workers. It didn't matter that oil companies already maintained their own living quarters—public money to build was there. (In a 1991 visit, I found the building vacant and apparently never used.)

But while the paper trail and witnesses had clearly established a 10 percent kickback pattern, for conviction under RICO the government had to prove that the municipality had been harmed. Spaan took no chances with the jury. He had hired Roger McAniff, an associate professor of engineering at the University of Alaska Anchorage, to do a project cost summary of contracts awarded to the inner circle principals on the North Slope. Working from a large chart that Spaan posted in front of the jury, McAniff analyzed twenty-three contracts involving nine projects, testifying that each was grossly overpriced. Blackstock had been paid $8,785,000 for its services in building health clinics, when a more realistic cost would have been $3,523,000. MMCW had been paid $24,435,000 for consulting, when McAniff's analysis showed that it had done about $13,200,000 worth of work. A firm named Olympic faithfully kicked back 10 percent of the $25,639,000 it was paid to build fire stations that could have been built for less than half that amount.

The defense attorneys pounced on McAniff. They grilled him for two months, attempting to show that he had greatly underestimated the cost of building on arctic permafrost. They cited the extraordinary costs of the high school and Utilidor, which had been started by Brower's predecessor. But while the defense attorneys parried with the witness, jurors were studying the project cost summary chart in front of them. Several could be observed doing their own math.

Finally, the government opened the subject of the defendants' personal wealth. Now jurors saw figures they didn't need a math course to understand. FBI agent Talbert, an accountant by training, calculated Dischner's net worth at $150,000 at the start of 1981. By 1984, he said, Dischner's net worth rose to between $11 million and $12 million and included properties in Anchorage and Palm Springs. The government introduced Dischner's income tax records, over strong objections by the defense attorneys. The returns showed that Dischner had listed the numerous companies associated with the North Slope constructions as "clients." The largest was H.W. Blackstock. In one year alone, Blackstock paid $1.6 million in fees to Trust Consultants, a business solely owned by Dischner. The government maintained that Dischner had not only failed to pay proper income taxes in 1981 and 1982, but had filed no tax returns at all in 1983 and 1984; his estimated income for those two years was $19 million.

In May 1989, eight months after it started, the longest and most expensive criminal trial in Alaska's history came to an end. More than one hundred witnesses, including nearly everyone who had any connection with the government in Barrow, had been called by the prosecutors. Thousands of documents and other items introduced as evidence filled the courtroom. Neither Dischner nor Mathisen took the stand, and the court had seemed relieved when Mathisen's defense attorney called only one witness. By the end, the attorneys, the jurors, and even the judge appeared on the edge of nervous exhaustion.

The jury deliberated for sixteen days. On May 23, it found Dischner, now seventy-three, and Mathisen, fifty-seven, guilty of more than twenty counts each of racketeering, fraud, bribery, and accepting kickbacks from contractors. U.S. district judge James M. Fitzgerald sentenced each man to seven years in federal prison and ordered them to forfeit more than $5 million in property.

But the euphoria in the U.S. attorney's office was short-lived. Exhausted, Spaan resigned, saying he needed a year off to refresh body and spirit with a trip around the world. An assistant, Mark Davis, was named acting U.S. attorney. It was up to him to direct the prosecution of the two major bribery cases that remained—against the MMCW engineers and the Seattle businessmen who ran the H. W. Blackstock Company, who together had profited more than anyone from the North Slope contracts.

The MMCW gang, architect Allen McDonald and engineers Peter White and David Coffman, had been indicted on thirty-nine counts of racketeering, bribery, and tax fraud. If convicted, they faced more than twenty years in prison. Their trial was initially scheduled for Fairbanks, but no judge was available because federal courts were too busy and recent retirements had drained the judicial ranks. The search for a federal judge elsewhere took on comic proportions.

First, the trial was moved to Los Angeles. The prosecutors in Alaska protested. Then it was moved back to Fairbanks, but the Los Angeles judge, a onetime amateur ski champion, balked because, among other reasons, he considered the skiing inferior in Alaska. The case was finally assigned to federal court in Portland, Oregon, where it simply evaporated in March 1990. Despite protests from his young assistant, Neil Evans, assistant U.S. attorney Stephen Cooper from the Fairbanks office negotiated and signed an appalling plea bargain.

Cooper agreed to reduce the thirty-nine-count indictment against the MMCW defendants to a single felony with no jail time, no fine, no community service, and, worst of all, no commitment to testify in the Rogstad corruption trial yet to come. In comparison, back in 1987, Geoffrey Fowler, a fourth member of the MMCW group who had agreed to testify for the government, was sentenced to six months in prison, a $10,000 fine, and one thousand hours of community service.

Acting U.S. attorney Davis claimed he had no idea such a plea bargain was being signed and said it was done without his approval. He sped to Portland in an attempt to get the presiding judge, James Burns, to rescind it. Judge Burns was incensed. He castigated Davis for "this thoroughly unedifying display" and threatened to reconsider the constitutionality of all the charges. The plea bargain stood, and a chastened Davis returned to Anchorage. Mike Spaan, back from his vacation and now in private practice, called the whole matter "unconscionable." Neil Evans, the assistant, resigned in protest. Cooper did not respond to media queries.

The third and last of the three major racketeering cases was placed in the hands of Peter Gamache, who was on loan from the state attorney general's office. In February 1990, Gamache had successfully indicted the two Seattle men who ran the H. W. Blackstock Company. Kenneth Rogstad, the president, and Wayne Larkin, the director of administration, were charged with paying more than $2.5 million in bribes and kickbacks to obtain $140 million worth of slope business during the Brower administration. If convicted, each faced twenty years in prison and the forfeiture of millions of dollars.

The case had more than usual media interest because Rogstad was a politically prominent former chairman of the Republican party in King County, which encompasses Seattle. Newspapers reported that he had connections to the White House through a Blackstock director, Seattle lawyer James Munn, who had run Ronald Reagan's 1980 presidential campaign in Washington State.

The case began to unravel before it began. Right off, the Seattle pair succeeded in moving the trial to federal court in Tacoma, Washington, roughly 3,000 miles from Barrow. Next, Peter Gamache unexpectedly resigned to take a state job in Kodiak. His departure further crippled the U.S. attorney's office. Almost by default, the case fell into the lap of Karen Loeffler. With barely a month to prepare, she would have to take on two nationally prominent defense lawyers—noted San Francisco attorney Marc Topel, representing Rogstad, and Chicago attorney Thomas Decker, retained by Larkin. Both had had plenty of time to do their homework.

In September 1990, Karen Loeffler packed up three thousand pounds of documents and headed for the U.S. district court in Tacoma. A senior judge, Jack Tanner, had been assigned to the case. Karen soon discovered that Tanner, a retired jurist, was also an impatient judge. "The government's case depended upon building a forest of evidence as it did in the Dischner and Mathisen cases," remarked Hal Spencer, the Associated Press court reporter who covered the trial. "But every time Karen tried to produce a tree, the judge or the defense knocked it down."

Meanwhile, the government's Eskimo witnesses, Brower and Irving Igtanloc, were torn apart by the defense. Yes, they had received items of value ranging from $37,000 boats to groceries and free housing from Blackstock. No, they conceded, these were not bribes, only gifts. Yes, they had been billed for these items by Blackstock—after the government started its investigation in 1984. No, they had never paid. Several of the more experienced witnesses who might have helped, such as borough lawyer Chris Mello or Roger McAniff, were never called to the stand at all.

"There is no question that the mayor and the public works commissioner were bribed," Judge Burns commented after the prosecution rested its case. But, he said, whether Rogstad and Larkin had been involved remained at issue. Topel and Decker quickly built a wall of doubt for the jury. First, they argued that Blackstock had been paying fees to Dischner for years in return for consulting services. What Dischner did with the fees was of no concern to them. As for the items the company had given to Mayor Brower, those were not favors, they stressed, but items he was expected to pay for. It happened that Brower was something of a deadbeat, Topel declared, but Blackstock, however belatedly, had billed him for the boats and housing.

The trial of Dischner and Mathisen, which involved similar allegations, had lasted eight months. This one lasted two weeks. The jury acquitted both defendants. Afterward, a juror remarked, "We felt something was fishy, but the government never proved it."

It was a sad ending to a long and courageous investigation. FBI agent Bruce Talbert had devoted six years of intense work to the case and paid for it with a broken marriage. Karen Loeffler is now saddled with responding to eternal appeals as Dischner and Mathisen, among others, fight going to jail. She still manages to be cheerful, although she feels she was betrayed by a lower forty-eight jury with an ingrained perception that such doings were just normal for Alaska.

More than a dozen North Slope officials, contractors, and consultants were convicted. The Eskimo community still hopes to recover some of the splurged funds through a maze of civil suits that were still slowly moving through the Anchorage courts in 1992. But the only sure winners will be the lawyers.

Aside from Rogstad and Larkin, only one other defendant, Brower's lawyer, Marty Farrell, was acquitted. Dischner and Mathisen were ordered to begin serving their jail terms on March 2, 1993. Until then, the only ones to serve time were three Eskimos. Ex-mayor Brower, sobbing before the court, was sentenced to thirty days and three years' probation for tax evasion. Irving Igtanloc did six months in a halfway house for bribery. His brother, Myron, spent six months in a minimum-security prison in Washington State for tax evasion.

At best it was a cloudy victory that left Alaskans wondering why neither the U.S. Justice Department nor the state congressional delegation had helped arm the U.S. attorney's office to see the battle through.

CHAPTER 11
Anchorage Meets Dallas

LIFE TURNED SOUR in 1982 for Rick Chiappone, a hustling, thirty-three-year-old paperhanger who had been doing well covering the walls of glitzy Las Vegas hotels and casinos. His marriage had broken up, he was under court order to pay off all of the leftover house bills, and his girlfriend was egging him on to make a clean break from Vegas.

So Rick packed up his tools, married his girlfriend, and headed for Alaska to start a new life. He had read that Anchorage was one of the ten fastest-growing American cities, and it lived up to its billing. "There were cranes and skeletons of high rises on the horizon, a bulldozer on every corner, and pickup trucks with Idaho, California, and Washington license plates at the construction sites," Rick says. "The roads were jammed with drivers who seemed impatient to get to where they were going."

Rick spent $1,500 that he'd salvaged from the bill collectors on a Chevy pickup. He painted a sign on the side that read, "Paper Hanging." Then he put down his last dollar to install a telephone in a tiny one-room apartment he felt lucky to rent at $450 a month.

"I just walked into places and started bidding, and the phone started ringing as soon as people heard there was another paperhanger in town," Rick says. "I stepped into a pot of gold. Public buildings and private houses were going up everywhere and anyone with any skill in construction could get a job. Up here people were geared to hanging paper only in bathrooms and kitchens. No one was geared to do the fast-paced wall installations I did in Las Vegas. Soon, my wife and I had all the business we could handle. She has a master's degree in anthropology and eight years' experience as a social worker, but she worked with me, covering walls. Our gross receipts hit two hundred twenty-five thousand dollars in two years, and I even got in some fishing."

Rick's success story was not unusual in Anchorage in the early 1980s. Unlike Barrow, which boomed from property taxes on oil facilities at Prudhoe, or Valdez, which prospered from the big pipeline payrolls and terminal revenues, Anchorage thrived on a later wave of prosperity that oil discovery brought to Alaska.

Through the hard campaigning of city fathers, with *Anchorage Times* publisher Bob Atwood leading the way, the oil companies were persuaded to locate their Alaska headquarters in Anchorage. Fairbanks, which was much closer to the Prudhoe oil fields, was the logical site, but, as Atwood says, "we were much more inviting." Anchorage civic leaders promised greater cultural opportunities and made much of facilities at the city's all-weather international airport. (The frequency and severity of Fairbanks's winter ice fogs was well known.) Meanwhile, Atwood says, Fairbanks business leaders were raising rents and otherwise preparing "to screw the oil companies."

The oil companies helped create the first cosmopolitan city in Alaska. In 1983, Atlantic Richfield (ARCO) finished a twenty-one-story office tower of glass and steel in downtown Anchorage at a cost of $65 million. Sohio, BP's domestic subsidiary, matched that with a handsomely landscaped fifteen-story midtown edifice costing $75 million. Nelson Bunker Hunt invested $45 million of Texas money in a twenty-story high rise within easy walking distance of ARCO. Lesser high rises sprouted all over the city. Acres of tiny cottages were bulldozed to make way for a forest of gleaming office buildings.

A few spirited citizens tried to persuade the city administration to consider saving a block of the older homes out of respect for heritage. A public hearing was set, but it became academic when one developer learned of it and moved in at dawn to bulldoze three historic homes in the block. ARCO did spare two frontier-era homes on its site and moved them at its expense to the city dump until Historic Anchorage Inc., a civic group formed to preserve the frontier legacy, could decide what to do with them. But the preservation movement lost steam when the city decided that private funds would have to take it from there.

By 1982, oil royalties and commissions from Prudhoe Bay pushed state revenues to $4.5 billion a year. The state spread the money around lavishly—not only for its own projects, but also as direct cash payments to municipalities. The distributions were based on a formula of $1,000 per resident, to be used any way the towns saw fit.

With about half the state's nearly five hundred thousand people, Anchorage found itself awash in money. True to its promise to the oil companies, it earmarked more than $250 million for the cultural amenities that would make it a world-class city—theaters, concert halls, libraries, and museums. The centerpiece of the undertaking, called "Project 80s," was the $75 million Alaska Center for the Performing Arts, which houses three state-of-the-art theaters. Critics of its design—of which there still are many—say the building resembles a "gigantic squatting toad." Nevertheless, star performers, from Jay Leno to the San Francisco Opera, regularly grace its stages, often with praise for the facility.

The George M. Sullivan Sports Arena ($34.5 million) seats eight thousand and can accommodate events from ice hockey tournaments to rock concerts. The majestic Loussac Library building ($41.6 million) is a versatile edifice, though some contend that more should have been spent on books and less on its elaborate Roman design and fountains. The Museum of History and Art ($25.5 million) became an important anchor on the edge of the refurbished downtown and a major visitor attraction. The Egan Civic and Convention Center ($26.8 million), across the street from the performing arts center, offers a convenient location that draws conferences.

The arts, which had always been lively in Alaska if largely homespun, became adventurous, polished, and exceptionally well heeled. Christine D'Arcy, who heads the Alaska State Council on the Arts, an allocating agency, says, "The biggest legacy of the oil money was the development of cultural facilities throughout the state. But what also happened was that the arts organizations were able to undertake sustained creative development and hire staffs to promote funding. As a result, today we have a twenty-three-million-dollar arts industry that provides income to four thousand people throughout the state."

In 1967, the newly created arts council's budget was $54,778; fifteen years later, oil revenues shot that up to $6.2 million, the highest per capita subsidy in the nation. Arts activities bloomed throughout the state, attracting artists from throughout the country. Anchorage, in particular, astonished visitors with its resident opera, symphony, and concert associations and especially its excellent repertory theater.

The cultural vigor of Anchorage undoubtedly made the harsh climate of Alaska more inviting to oil executives and other employees offered transfers from Texas and other southern states. ARCO and BP, the major Prudhoe partners, added substantial inducements of their own. ARCO, for example, gave bonuses of 45 percent on the first $24,000 of base salary and 25 percent on the remainder. Each adult received $1,300 toward air travel, and each child $1,000.

The influx of so many families from sophisticated Dallas and other oil-rich cities further rearranged the demographics of Alaska. Soon, blocks of elegant homes and condos appeared, some alongside the not-so-elegant jerry-built apartment houses, tar-paper shacks, and trailer courts that housed an earlier generation of Alaskans. Nordstrom, the Seattle-based chain of tony specialty shops, renovated the store it had bought in 1975 in anticipation of the affluent customers oil would provide. Boutiques, interior decorating shops, and trendy restaurants mushroomed where scrubby brush had stood. Workers who headed downtown over gravel in the morning were startled to find themselves driving home over freshly paved roads at night. The city's pace was frantic.

In 1983 alone, permits were issued for a billion dollars' worth of construction, exceeding the combined building permits issued that same year in the also-thriving cities of Seattle, Portland, and Honolulu. Bankers, from inside the state and out, stormed into the city as willing lenders, taking full advantage of state programs that guaranteed loans. Easy borrowing further fueled the boom.

Given such volatility, the area was rife with opportunities for immense private profiteering. It was inevitable that the action would attract the likes of Peter Zamarello, an Alaskan entrepreneur without parallel, who in a few brief years would rise from rags to become the richest man in Alaska.

Zamarello was born in 1927 to poor Italian parents on the Greek island of Cephalonia. His formal education ended in elementary school. He says he bribed a bureaucrat to get him a high school diploma so he could join the merchant marine. When his Greek freighter reached Albany, New York, in 1953, Zamarello jumped ship, headed for New York City, and within forty-two days married a Polish girl who had American citizenship. (When asked how a poor boy fleeing immigration authorities could sweep an American citizen off her feet in such a short time, Zamarello replied, "Don't you know the smartest people in the world come from Cephalonia? That's spelled C-e-p-h-a-l-o-n-i-a.")

His timely match kept the immigration authorities off his back long enough for him to work his way across the country as a carpenter until he reached his goal—Alaska. Zamarello's father was one of the immigrants who'd come to the tent city of Anchorage in 1915 to build the Alaska railroad, and he'd constantly talked of the wonders and potential of the land after his return home in 1924.

Anchorage in the 1960s was teeming with hustlers like Zamarello, many of them busy combing the oil scouting tip sheets for where the big plungers were leasing land. While the oil game tempted him, Zamarello soon saw a better way to make his fortune in Alaska. He wielded his hammer only as long as it took him to save enough money for a down payment on a loan to buy a parcel of land. He quickly discovered the weak underbelly of the banking bureaucracy. Credit checks in Alaska were so loose that when the money from one bank loan ran out, he found he could go to another bank and get another loan.

Zamarello's first construction loan—$150,000 in 1969—enabled him to build a shopping center in Muldoon, a shabby but strategically located neighborhood near the gates of the huge Elmendorf Air Force Base. Pizza parlors, convenience stores, and other businesses were quick to rent when they noticed the steady traffic. Next,

Zamarello bought a 568-unit trailer park. Convinced that imminent oil discovery would make the city boom, he used his new flow of revenue to buy cheap land around Anchorage. When pipeline construction started in the 1970s, he opened an office in Beirut, Lebanon, where he extolled the potential of Alaska to oil-rich Arabs, and, he says, successfully sold Alaska land to wealthy sheiks. (His landlord in Lebanon, he claims, was Jaber al-Ahmed al-Sabah, who became the Kuwait emir deposed by the Iraqis in 1990. In an interview at his Anchorage office in February 1991, Zamarello fumed over U.S. involvement in the Gulf War. "That Jaber Whatever-the-Hell-the-Rest-of-His-Name-Is had fifty concubines and two wives. Still, he was always trying to screw my secretary. And now we have the whole American army over there trying to save a guy like that.")

When the oil companies opted to build their Alaska headquarters in Anchorage, land values took off like a missile. Zamarello was perfectly positioned. His real estate leverage propelled him out front in the frenzy of development that transformed Anchorage and much of its environs. He borrowed, built, and borrowed some more—at least a billion dollars, by his own estimates—to create an empire of dozens of shopping centers, malls, office plazas, condos, and trailer parks. Banks in Alaska and Washington State courted him, rarely questioning his single-page, unaudited statement of financial worth. By 1985, Zamarello had become the largest and richest developer in the state. Known as "king of the strip malls," he employed a thousand workers. He could even boast that he owned city hall; he'd bought the building Anchorage leased for its principal municipal offices with an $8 million loan in 1984.

His personal fortune grew to a reported $292 million. He freely accepted the title of wealthiest man in Alaska—and did his best to live up to it. He lived in a $1.2 million mansion in Anchorage, bought condos for himself and his wife in Tacoma, and owned an $850,000 retreat in Honolulu. He invested millions in four financial institutions, acquired $1.4 million in Krugerrands, and collected a trove of art, expensive furniture, and jewelry valued at $11 million. His favorite bauble was an $80,000 gold-and-diamond watch. The only other one like it was said to belong to King Fahd of Saudi Arabia. Although a self-described atheist, he paid a large share of the bill for a $3 million cathedral that became the headquarters of the Orthodox diocese of Alaska. ("I did it to honor my parents.") A grade-school dropout, he pledged 40 acres of lakefront property and cash valued at $480,000 to Alaska Pacific University. ("I want others to have what I missed.") He made a major contribution to build a mobile dental clinic for children in Greece. (The Greek government awarded him its medal of honor.)

Anchorage has paid dearly for Zamarello's rise to riches. The stamp of banality he put on the city has persisted. His dozens of strip malls and commercial centers, some with hazardous traffic entrances and many in incompatible neighborhoods, stand as testaments to the zoning and planning standards that were ignored while Anchorage boomed. Most of his malls and plazas are undistinguished rows of humdrum storefronts, strung together under elongated blue roofs. He threw them up wherever he chose, even in a suburb 40 miles from Anchorage, and because the city did not check the crazy-quilt pattern of its commercial development until it was too late, it is fated to live with them.

Zamarello refers to "the pygmy minds of people who run government" and boasts he paid thousands to insulate himself from "stupid" public officials who were in a position to challenge him. One day, he regaled the students in a University of Alaska Anchorage economics class with stories of his bribes and laundered campaign contributions to governors, mayors, and other public officials. In 1977, during the administration of Anchorage mayor George Sullivan, Zamarello says he even bribed two greedy building department employees by check. "I complained to the Anchorage assembly. They investigated. The ethics chairman said the inspectors used poor judgment. The two guys get fifteen-day suspensions—with pay," Zamarello says, with a contemptuous laugh.

In 1982, Tony Knowles, a youthful, Yale-educated native of Oklahoma, took over the mayor's office after a hard-fought victory, defeating a prodevelopment legislative leader backed by aggressive real estate interests. A biker, cross-country skier, and lover of Alaska's open spaces, Knowles campaigned on a promise to bring Anchorage's runaway building under control. The new mayor promptly cleaned out the zoning and planning boards, which had long been dominated by real estate people. Then he put across twenty-one revisions to the planning code, ranging from a policy to protect and, when possible, acquire wetlands to regional controls on strip zoning.

It was only a matter of time before Zamarello and Knowles collided. No sooner had the new mayor put his development restrictions in place than Zamarello announced that he intended to build a thousand-room hotel at the intersection of Minnesota Drive and Raspberry Road, which is a gateway to the international airport and on an easy access route to downtown. But the site was smack in the administration's newly designated wetlands area, and the city refused to give Zamarello a permit. The standoff intensified when a city employee inspecting the site found a duck nesting on a batch of eggs.

Zamarello responded in his usual flamboyant style. He rented a room in the name of "A. Duck" at the Captain Cook, the city's five-star hotel. Then he sent a copy of the reservation to city hall with an offer to house the "fuckin' duck" in the hotel room at his own expense until the eggs hatched. The Knowles administration would not be moved. Stunned by the rebuff, Zamarello vowed he would get even with the bird lover in city hall. (He waited until 1990, when Knowles ran for governor, to exact his revenge. As the campaign heated up, ethics became a pressing issue, whereupon Zamarello surfaced in the media, claiming that he had illegally laundered $27,000 in contributions to Knowles's 1984 campaign with the mayor's full knowledge. Knowles admitted receiving consecutively numbered cashier's checks in the names of Zamarello and his various employees but said he was unaware the contributions were illegal. He concedes that the allegations hurt him in the race, which he lost.)

By late 1985, university economists were warning that Anchorage was heading for a building glut. Some developers pulled back, but Zamarello declined to slow down. Inevitably, laws of economics kicked in and Zamarello's empire collapsed almost as spectacularly as it began. "The Anchorage area simply became overbuilt, and bankers permitted the overbuilding because they misread the nature of the boom," Arnold Espe, a surviving banker, recalls. "Too many buildings, too many barbers, too many everything. The boom never developed the underpinnings of permanent private employment. Population grew as long as government kept throwing massive amounts of money at virtually everything. When people saw that government no longer had the money to throw, they left the state almost as fast as they came in."

The steep dive in the price of oil early in 1986 hastened the exodus. Alaska crude fell from $27 a barrel to less than $10, which translated into a loss of more than $1 billion in already budgeted revenues in the space of six months. Governor Sheffield declared a fiscal emergency. He cut nine thousand jobs from the state payroll. "It was as though you lived in Detroit and Detroit closed its automobile plants," recalls Scott Hawkins, who runs the city's economic development corporation. "The immediate impact was less cash on the street, fewer people buying goods, and stores and businesses suddenly struggling to stay alive."

Yet even as other businesses started retrenching, Zamarello disclosed plans for an eighteen-story Anchorage bank building. He was particularly pleased by the location, he said, because his penthouse would blot out the view of the Chugach Mountains for the National Bank of Alaska across the street, the only bank in town that refused to lend him money.

But by 1986, Zamarello's store vacancy rates had increased drastically and his newest malls had no takers. Newspapers later reported that Zamarello's lease income dropped from about $13.2 million to $7.2 million in that one year. The banks that had lent Zamarello money on the basis of that one-page unaudited statement now began to look deeper. They soon found his business was in distress and worsening by the day as Alaska real estate began a freefall. When a contractor sued to collect payment due and won a $4.3 million judgment against Zamarello, his empire was pushed over the edge. In August 1986, he filed for protection under Chapter 11 of the bankruptcy laws. The depositions revealed that Zamarello was swamped by debts estimated at about $150 million and pursued by about 250 angry creditors.

Before going broke, Zamarallo built 4 million square feet of retail space, which amounts to more than half of the retail footage in the state. Now he handed the state one more legacy—the most complicated bankruptcy case Alaska had ever encountered. Court-appointed examiners tried to untangle his seven interlocking companies and attempted to question Zamarello on the whereabouts of his fortune. It was tantamount to trying to revive memory in a man with a serious case of amnesia. He turned the bankruptcy proceedings into a comic skit so intriguing that even *The Wall Street Journal* devoted a front-page story to it.

Questioned about a rumor that his estranged (but still friendly) Polish wife, Patricia Zamarello, carried a suitcase containing $20 million out of the country and deposited the money in a Swiss bank, Zamarello laughed. "It's hard to get even one million dollars in a suitcase—I know, I tried it," he said. "If she could carry twenty million dollars, I'd put her in the weight-lifting Olympics." Had he transferred any possessions to his wife? Yes, Zamarello said, he did give a portion of his assets to his wife in exchange for other property in Beirut. "Don't go look for it," he told the examiners. "It's not there anymore. They blew it up." He claimed he had no job, no income, and no cash except the $8 or $10 on his person. Turning to the lawyer who was taking his deposition, he asked, "You want it?" At another point, he said he'd "rather be a pimp with a purple hat and a feather" than be associated with banking. After the bankruptcy was over, he added, he would apologize—to the pimps.

Zamarello's crash was one of the first in the severe recession that gripped Alaska in 1986, and the hardest. The tremors of his collapse were felt by banks, suppliers, and the entire real estate industry. The fifteen-year mortgages with variable interest rates many of the banks had stipulated became too stringent as the economy began to soften. Landlords unable to fill their units simply declared bankruptcy and left lending agencies to deal with the problem. Nine of Alaska's fifteen banks went broke or were forced into mergers soon after, most of them overleveraged in real estate and many of them carrying uncollectible loans made to Zamarello.

John Shively, who became chief executive of United Bancorporation, Alaska, Inc., after leaving Governor Sheffield's administrative staff, was among those who had the unpleasant task of informing stockholders their bank was going out of business. "We lost $83.7 million in the first six months of this year," Shively wrote in 1987. "That is a pittance compared to the billions of dollars the big boys are losing. . . . The majority of the people felt the real estate development boom would last forever . . . but Alaska banks were financing the building of condos, houses, and shopping centers for the people who were building condos, houses, and shopping centers. When the building stopped and other sectors of the economy began to disintegrate, people working in the construction industry and other ailing businesses began to leave and now we don't need all those condos, houses, and shopping centers."

The Federal Deposit Insurance Corporation (FDIC) took over $1.5 billion in bad loans and defaulted property from the failed Alaska banks and thrifts. On a per capita basis, the government bailout rivals the national savings and loan scandal that would follow. In an attempt to recoup losses, the FDIC sued thirty-three former bank officers and directors in 1991 for $55 million. Charged with gross mismanagement, breach of duty, and reckless real estate lending, most of the defendants are pillars of the Alaska establishment. Among them were three top officials of Governor Hickel's administration, including his commissioner of revenue; former and current legislators; Native corporation leaders; and even the assistant publisher of *The Anchorage Times.*

While banks were the most spectacular failures, hundreds of other trades, professions, and individuals also flooded the bankruptcy courts. Employees were left stranded as businesses closed their doors. The carpetbaggers who survived were those who had other resources to fall back on. Among them was Rick Chiappone, the hustling paperhanger. His well-educated wife managed to get a job teaching school and her income paid the bills while they rode out the recession. The stable incomes of oil industry executives and office staffers kept Anchorage from flat-out collapse; the monetary excesses of those who had come just to cash in were already gone.

After three years in bankruptcy, Zamarello was back on the scene in 1989. His complex reorganization left him with only a fraction of his holdings. Most of his malls were becoming eyesores, with flaking paint and neglected landscaping. Unfortunately, the pattern he had set had inspired other developers to produce many more miles of urban blight. Zamarello, meanwhile, rejoined the business world, swearing that he would never build anything in Alaska again, "not even my tomb." Some contend that he must nevertheless be considered one of the lucky ones. While he lost an empire, what did he really lose? As one bitter banker observed, "Pete could have fared a lot worse. He could have lost his own money."

CHAPTER 12
The Iron Fist Of Jesse Carr

RON WINDELER, a thin, wispy-bearded son of a homesteading family, used to hear stories of ruthless characters in Alaska while growing up on the remote, moose-trampled acres of the Kenai Peninsula. But he never really knew the meaning of ruthless until he left the failing farm and met Jesse L. Carr, the boss of the International Brotherhood of Teamsters union in Alaska.

In 1972, Windeler, then twenty-three, got a job as a technician on the White Alice, an RCA/Alascom project that installed communications systems for radar stations on the mountaintops along the Bering Sea in Alaska. He had heard of Carr, the state's fastest-rising labor leader, and even benefited from his clout during his first year on the job. The aggressive Teamsters had just ousted the sleepy International Brotherhood of Electrical Workers (IBEW), and Carr promptly got the technicians a 41 percent raise—from $4.36 an hour to $6.16. And that was the year President Nixon had slapped a 5½ percent wage freeze on the nation.

Windeler was impressed, but he did not see the man in action until 1977, when he attended a Teamsters union meeting in Fairbanks at which Carr was presiding. He watched in awe as the burly, flushed man, alternating between bombastic invective and studied silence, kept a roomful of normally rowdy Teamsters in docile decorum as he discussed mounting an assault on Alaska's antiunion bastions.

The principal piece of business that required membership action was ratification of the new agreement with Alascom, the statewide phone company. The pact covered Windeler and the other technicians working on White Alice and Carr said he had pounded out a good contract. Windeler had a question, so he did what he thought any concerned union member would do—he raised his hand. Carr ignored him and went on. "We'll ratify by a voice vote," he announced.

Windeler rose to his feet. "I move we vote by secret ballot," he said. Carr scowled, slammed his gavel, and angrily ruled Windeler out of order. The members gave Carr the chorus of *ayes* he was waiting for. After the meeting, Carr approached Windeler. "How's it going, son?" he asked. Then he poked a finger in Windeler's belly and walked off.

When Windeler reached home that night, he noticed his car had a flat tire. The next morning he saw that a screw had been driven into the sidewall. The tire had held air while the car was parked, but the friction opened a leak as soon as the car began moving. "Someone skilled in inflicting damage did it," Windeler says. "You can fix a puncture in the tread of a radial tire, but when the side is pierced, the tire becomes unpatchable."

Windeler became all the more determined to make his presence felt at union meetings. He studied the bylaws and Robert's Rules of Order. He talked to members about defending their rights. He challenged procedural points from the floor. And over the next six months he had thirteen more flat tires, all the result of screws or nails driven into the sidewalls.

Windeler asked Fairbanks police for help. The police took notes, but nothing happened. Then he talked to reporters at the local daily, the *Fairbanks News-Miner*, and the paper obtained a tape recording of a night of Carr's dictatorial rantings, printing several of the more colorful quotes. At the next meeting, Windeler encountered two thugs in a hallway. One grabbed him from behind and crushed him in a bear hug while the other snatched his briefcase, scattering the contents on the floor. He had no tape recorder, but they sent him sprawling across the hall anyway. By now Windeler had learned a basic lesson: Jesse Carr ran a rigid union.

A high school dropout and ex-Marine, Carr learned the ropes under Southern California Teamster boss Dave Beck, who became president of the international Teamsters and eventually landed in jail. Carr came to Alaska in 1951 and drove a gravel truck for two years. In 1953, at age twenty-seven, he became business agent for the infant Local 959, which had about five hundred members and was several hundred dollars in debt. Seven years later, Carr persuaded the international union to merge all four Teamster locals in Alaska and put him in charge.

Oil explorations at Swanson River and the Cook Inlet had brought in new business and therefore new opportunities for organized labor in south-central Alaska. Carr did whatever he thought was necessary for his union's share. By the early 1970s, he had been hit with six felony arrests, on charges ranging from extortion to embezzlement. He beat every one. He was acquitted on four of the counts and the other two were dropped when, according to court records, "the principal witness had to undergo extensive medical treatment and was in no physical condition to testify at the trial." No explanation of what caused the disability was given to the media.

The aura of invulnerability added to Carr's growing prestige inside and outside the union. He became known as a master of intimidation. By the mid-1970s, the Teamsters in Alaska held an estimated five hundred contracts, many of them wrenched from management by their leader. Few companies dared to say anything publicly about Carr's tactics, but privately, they reported, he became a beast in contract showdowns. William Heintz, a Teamster, says Carr once drop-kicked a rubbish barrel through a window during negotiations with the trucking company Sea-Land Service.

Besides organizing the traditional truck drivers and haulers, the Teamsters became the bargaining unit for such unlikely groups as butchers, bakers, bank clerks, high school principals, and even the Anchorage police department. Every new member swelled the union treasury, and, Carr, as secretary-treasurer, controlled the purse strings.

By the time pipeline construction peaked in the middle 1970s, Jesse Carr had already become one of the most feared men in Alaska.

Teamster membership in the state soared to twenty-three thousand, making the union the largest and most influential special-interest body in Alaska. Carr cowed not only employers but important public officials as well. Former Alaska commissioner of labor Lew Dischner lobbied the legislature on behalf of the union and dispensed favors to the right people. Teamster money was distributed generously at election time, and candidates from both political parties vied for Teamster endorsements. Newspapers began to refer to Jesse Carr as "the most powerful man in Alaska . . . his word [is] capable of shutting down the state."

Rumors about Teamster ties to organized crime surfaced during the building of the pipeline in 1976 after two of Carr's people working at the main warehouse in Fairbanks were abducted and murdered, but no one ever investigated. "One of my biggest regrets is having been too green to do what needed to be done in unraveling the double homicide associated with the Teamsters union and the warehouse terminal in Fairbanks," Dan Hickey, then state prosecutor, says. "Around this time, a union steward talked the parole board into freeing a man who was serving time for a botched paid hit. He is paroled and two people at the terminal disappear."

An FBI strike force from California conducted the only grand-jury investigation into pipeline racketeering, but it concentrated on prostitution in Valdez rather than violence in the Teamsters union. Six Alaskans were indicted in 1976, but the closest the FBI came to Carr was arresting his former lawyer. He was subsequently acquitted, as were all five other defendants.

During pipeline construction, about $1 million a week flowed into Local 959's trust accounts, nearly all of which Carr dominated as an overpowering trustee. The largest of those entities was the pension fund, which collected $220 million in mandatory employer contributions by the time the pipeline was finished. It was evident that many workers paying into the pension fund were Teamsters only for the duration of the pipeline project. They were unlikely to stay in Alaska for the ten years stipulated for vesting. Jesse Carr saw the opportunities in those excess millions. He set out to literally build an empire.

In 1974, Congress had passed the Employee Retirement Income Security Act (ERISA), a long-sought curb on the extravagance of union leaders. It was designed specifically to protect the nation's twenty-three million private, nongovernmental workers from having their pension funds squandered on dubious real estate deals, speculative stocks, or questionable loans. ERISA forbids trustees to invest pension funds to benefit parties of interest, or to make imprudent or self-serving investments, or to surpass specified limits of investments in any one geographical area.

Carr trampled every key provision of ERISA in putting together his Teamster colossus. He organized a nonprofit subsidiary called the Teamsters Local 959 Building Corporation and went on a construction spree. He got Teamsters to buy building bonds at $50 apiece, and banks courted the business. Even the conservative National Bank of Alaska readily provided construction mortgages, comfortably aware that the union was backed by an overflowing pension fund.

At a cost of $32 million, the Teamsters erected the two-hundred-bed Alaska Hospital along the south bank of Merrill Field, one of the first sights visitors see driving into Anchorage from the east. Union members would receive free care here, from womb to tomb, and Local 959 would control the hospital. It would hire administrators who would hire doctors willing to negotiate fair fees. For $8 million more, a 4,000-square-foot professional building was built to house a medical center. It was connected to the hospital by a $6.5 million mall especially for Teamsters, complete with pharmacy, shops, restaurants, bank branches, and a fountain. Ribbon-cutters called it "a monument to the growing strength of the Teamsters union."

The most modern recreation center Alaska had ever seen went up next, alongside the new seven-story office complex that would be named the Jesse L. Carr Building and house the relocated Teamsters headquarters. Union members had the use of an Olympic-size indoor pool, tennis courts, an indoor running track, and, most important, a well-equipped bodybuilding room—Carr and his chief lieutenants believed in lifting weights to maintain the Teamster image. A similarly well-equipped recreation center was built for members in Fairbanks.

Despite rebels like Ron Windeler, who was usually ruled out of order, Carr raised union dues from 40 cents a compensable hour to

50 cents so the union could expand its air fleet from two planes to four. It was true that Teamster business agents had more territory to cover. But it wasn't strictly business that accounted for Jesse Carr's frequent out-of-state trips these days.

Tapping the pension fund, he built a $120 million country club–condominium development in Indian Wells, California, 10 miles from Palm Springs, where the nation's wealthiest go to enjoy the prime desert climate. Named the Desert Horizons Country Club, the 275-acre complex sported a championship golf course, a lavishly equipped clubhouse, tennis courts, and swimming pools. The condos started at $300,000, in 1970s money.

Unfortunately, Carr pointed out, Alaska Teamsters, including himself, could not expect to receive special privileges or a price break at the resort because this would violate the ERISA regulations. He asked members to view the country club as a shrewd investment that would return great financial rewards to the union. Of course, he would need a Lear Jet for regular trips to oversee this money-maker.

Few Teamsters were fooled. Carr and his wife, Helen, a champion amateur women's golfer, were starting to lead a double life. While they lived modestly in Anchorage, they bought a well-appointed (and heavily guarded) condominium in an exclusive section of Indian Wells about a block away from Desert Horizons and within walking distance of the starting tee at the golf course. And while Carr drove a beat-up Monte Carlo sedan in Alaska, he tooled around California in a gold Lincoln Continental.

Carr's new upscale existence was not altogether inspired by a need for respite from Alaska's harsh weather. It was part of his grand empire-building strategy. Because of his excellent performance in Alaska, he had won a director's seat on the western conference of the international Teamsters, and he had visions of going higher.

The struggle for leadership of the international Teamsters union traditionally had been a tug-of-war between the eastern and central conferences, which had the largest memberships. But pension-fund scandals, Mafia dealings, and bribery convictions toppled one international union president after another. Dave Beck went to jail. Jimmy Hoffa served time and his mob-style disappearance afterward further stained the Teamster image. Roy Williams was president in the early 1980s, but even he was fighting accusations of mishandling pension funds and arranging kickbacks for Mafia figures. (He, too, would later go to jail.)

By Teamster standards, Carr was squeaky clean. The international chose its presidents at that time by vote of the conference delegates rather than of the members, and Carr cultivated a wide circle of contacts. His aspirations growing, he hosted Teamster power brokers at Desert Horizons and at his Indian Wells home. His guests led him to believe his performance and his jail-free record made him a strong compromise candidate for the highest union office. That expectation was reinforced by news reports that the U.S. Justice Department was becoming impatient with the succession of tainted presidents of the international and was even considering action under the racketeering laws.

But Carr's streak of luck started to change in the early 1980s. With the pipeline built, Teamsters began leaving Alaska in droves, drastically reducing union revenues. And with so much heavily mortgaged real estate, Carr's empire began to show cracks. An auditor's report filed with the U.S. Department of Labor showed that at the end of 1979 the Teamster building corporation had a deficit of $6.8 million and had lost more than $5 million in the past five years alone. Its biggest property, the Alaska Hospital, drained money at an alarming pace. Bankruptcy was inevitable if the pattern continued.

The burden of bailing out Carr's building program was falling on the Teamsters who chose to stay in Alaska, and many started to resent it. Ron Windeler was starting to have more comrades in opposition. In 1979, with barely enough supporters to fill a phone booth, Windeler had organized an opposition group called ROOR, which stood for Ruled Out of Order. The ROOR opposition slates never won a single union office, but they did drum hard on the theme that Carr was building a personal empire while denying democratic rights to the rank and file. The grievance was considerably strengthened by the tough and persistent reporting of the *Anchorage Daily News*. Then the city's struggling second paper, it had managed to keep a running account of Carr's excesses. A series on the Teamsters during the pipeline days won the paper and its courageous publisher, Kay Fanning, a well-deserved Pulitzer Prize for public service in 1976. (So feared was Carr in that era that no critical Teamster or employer allowed his name to be used for publication in that series.)

The growing dissatisfaction within the rank and file finally surfaced in 1981, when Carr attempted to bulldoze through one more contract at Anchorage Cold Storage, the state's largest wholesale distributor of meats, beverages, and grocery products. The firm's fiercely independent owner, Milt Odom, had battle scars from previous confrontations with Carr, and he decided that he wasn't going to take any more from the Teamster boss. He offered a take-it-or-leave-it deal. The Teamsters, at Carr's urging, turned it down and struck the company.

While Cold Storage workers picketed the plant, Carr called on other Teamsters to join in boycotting major grocery stores that sold Cold Storage products. Odom didn't flinch. The voluntary effort failed to produce the numbers Carr had expected, and it fizzled. Furious, Carr tried to up the ante. At a packed union meeting in Anchorage, he called for thousands of Teamsters from other crafts to join the biggest picket line Alaska had ever seen. He proposed that every Teamster picket four hours a week or pay a $100 fine.

Murmurs of dissent filled the hall. Ron Windeler, now working with Alascom in Anchorage, was among those who tried to find a floor microphone that night. "The place was so packed that some people had to sit in an overflow room, and they were unhappy about voting on something they couldn't hear, much less understand. None of us from the other crafts ever knew the terms in that contract the Cold Storage members turned down. And Jesse sandbagged the mikes so we couldn't ask. Only his people could speak. It was as though someone had written a white paper and assigned one paragraph to each staged speaker.

"When Bill Heintz, one of our Alascom techs, did manage to grab a mike and try to speak against it, he was ruled out of order and physically dragged out of the room. 'That guy [Heintz] is a little teched in the head,' a shop steward shouted. I finally got a chance to say it was wrong to force people on the picket line. Some others who got through offered an amendment to reduce the picketing requirement. It all added up to total confusion."

On the verge of losing control, Carr invoked his infallible parliamentary maneuver. "We will now put the proposal and amendment to a voice vote," he announced.

Windeler leapt to his feet and shouted that the mandatory picketing and the $100 fine were two different issues and each required a vote by secret ballot. But Carr was ready for him. He turned to a person beside him whom he described as a "parliamentary expert." The "expert" ruled that a voice vote was proper as Carr proposed. "A chorus of ayes was followed by an equal volume of nays," Windeler says. "I thought it was a dead heat, but Jesse hops up and says, 'The ayes have it.' And then he gives us a benediction and tells us to report to the picket line." Windeler, who had picketed voluntarily before the mandatory assessment, decided he would continue to picket. But he refused to sign the roster sheets that would legitimize the outcome of the meeting.

On December 22, 1981, ten Teamsters, with Windeler's name at the top of the list, filed a class action suit against Jesse Carr and Local 959 "to enforce their procedural rights." Each had contributed $50 to a kitty to start the fight. Carr was not worried. He was used to court battles, and he was confident that the deep pockets of the union would wear down the little band of dissidents until they ran out of money. But he could not so easily shrug off a different court problem that surfaced about the same time. For years, the U.S. Labor Department had been quietly investigating reports of Carr's manipulation of Teamster pension funds. Now a Department of Labor task force led by John A. LeMay, the Seattle-based administrator for labor management, had provided sufficient basis for the government to start a civil suit against Carr, the other Alaska Teamster pension trustees, and certain alleged illegal beneficiaries, one of which was Desert Horizons, Inc.

The Labor Department charged violations of ERISA in just about every major investment of Teamster pension funds in Alaska. The trustees had allegedly breached their fiduciary trust by financing the Alaska Hospital complex and by permitting about $18 million in unsecured loans to the hospital's related welfare plan, in violation of ERISA regulations that require money to be invested with some expectation of profit. In developing Desert Horizons, they allegedly violated the government regulations in three major ways: by using pension trust funds to acquire the property, by becoming directors of the country club and causing it to be built, and by causing the pension plan to bear the total risk for the project. The trustees further were alleged to have breached their trust by making "party in interest" loans totaling in excess of $8 million to companies regarded as friends of the Teamsters union. These were firms that granted lucrative contracts in exchange for substantial loans from the Teamsters pension fund.

The court did not address the issue of whether Carr and the other trustees would have to pay back millions of dollars in lost pension fund investments, and the full details of its decision were sealed in a consent order that was not released until many years later, so only the union's inner circle had any idea of the magnitude of Carr's trouble. But the union was ordered to begin disposing of the prohibited investments immediately. "For Sale" signs went up on the Alaska Hospital property, the adjacent medical center, and the mall. The union managed to sell the hospital within the year to the Humana Hospital chain for $60 million, thereby showing that it was abiding by the consent order in good faith.

But in the same year, Carr was hit with a devastating setback to his ego and, potentially, to the union's pocketbook. On December 2, 1982, Windeler's group won a clear first-round victory in federal court. U.S. district judge James von der Heydt ruled that Carr's picketing assessment in the Cold Storage strike had been passed illegally. He issued a temporary restraining order barring the union from levying assessments in the Cold Storage strike and certified the legitimacy of the class action suit, clearing the way for it to continue on the issue of damages. The injunction pulled a good number of Teamsters off the picket line. More crucially, it deflated the aura of Teamster infallibility. Angered by the illegitimacy of Carr's actions, Cold Storage workers voted several months later to decertify the union, thereby ending the Teamsters' jurisdiction and effectively breaking the strike. The ouster of the Teamsters at Cold Storage sent a signal to many other companies: You could stand up to Jesse Carr.

By 1983, the Teamsters local was still $4.2 million in debt, and membership had slipped to 12,000. But if the empire of the most powerful man in Alaska was crumbling, you wouldn't know it from Carr. "This local's never been in trouble. We've got lots of assets," he told *The Anchorage Times*. His drive for power in the international union continued. In 1984, he was elected head of the 450,000-

member Western Conference of Teamsters. Now he was potentially in line to succeed Teamster president Jackie Presser, who had been in trouble over his association with organized-crime figures from the day he assumed office. Alaska Teamsters say Carr's trips to California became more frequent and longer, leaving his confused lieutenants back home to wonder why the union empire kept losing money.

On January 5, 1985, Jesse Carr, age fifty-nine, was found dead on the living room floor of his Indian Wells home. His wife, Helen, found his body at 7:00 A.M. She said he had awakened at 3:00 A.M. complaining of pain, and told her he would take medicine and watch TV for a while. A coroner's inquiry attributed his death to heart failure.

The job of dismantling Carr's empire fell to Robert J. Sinnett and Jack Slama, two loyal business agents of Carr's. "That same January that Jesse died, Bobby and I sat down and started looking at our situation," Jack Slama says. "We saw we were losing an average of three hundred thousand dollars a month. The last bit of cash coming in was the five million dollars we received for the sale of the office building [to Humana Hospital]. That was eaten up so fast that on one occasion we were unable to meet the payroll."

They got more bad news from the courts. In the spring of 1986, a federal judge determined the union was liable for damages in the Windeler group's class action suit. The implication was tremendous. If every member forced to picket was to be compensated under the judge's formula, the Teamsters would have to pay out more than $10 million. In addition, an Alaska court ordered the local to pay five hundred thousand dollars more to a Fairbanks Teamster named Arlo Wells, who had filed his own suit over union threats when he refused to picket in the Cold Storage strike.

Having made the point that members suffered because of the lack of democratic process, Windeler says his group attempted to soften the financial blow to the troubled union. He said they offered to accept only a nominal settlement and to extend payments over six years. The final entry into the court record shows a judgment against the union of $4,898,000, plus $56,892 in costs and attorney fees.

It was the collapse of the real estate market in 1986 that proved the union's final undoing. The Teamsters Local 959 Building Corporation, the real estate subsidiary, had incurred heavy losses since its inception. It could find no buyers for even its most desirable remaining properties, notably the heavily mortgaged health clubs in Anchorage and Fairbanks. The union had no choice but to file for Chapter 11 bankruptcy reorganization before the year was out. "We came close to filing Chapter 7 [total dissolution], but we had too much pride to do it," Slama says.

The Alaska Teamsters emerged from bankruptcy court three years later with a plan that would allow the union to pay off about $9 million in debt over the next fifteen years. It did not need court prodding to liquidate the Desert Horizons condominiums and country club. "The country club investment and many of the other debts were a mystery to members," Jack Slama, who is currently secretary-treasurer of the union, says today. "But I firmly believe that Carr intended to pay off everyone. He was banking on becoming president of the international. It is not unusual for the international union to help out locals in financial trouble." Slama, who is as low-key as Carr was high-powered, has managed to reverse the financial losses by slimming staff, eliminating fringes such as the air fleet, and renegotiating the consent order with the Labor Department. ("Most of the charges were put on hold," he says.) But the Alaska union is a shadow of its former self. Its 1991 membership was estimated at seven thousand, a drop of 70 percent in fifteen years.

Many members left the state when construction stalled in the real estate bust. But just as damaging was the deflated image of the almighty Teamsters after Carr's setbacks in court and the Cold Storage showdown. The major oil companies opted out of contracts and continue to shun the Teamsters. Even traditional constituents such as truckers and warehouse workers are among those who have decertified the union. The decline of the Alaska Teamsters paralleled the decline of the union in the rest of the country; Carr's hunger for power and his rash, unrestrained expansionism are the corollary to the union's flagrant corruption and strong ties to organized crime on the national level.

The exodus of the Teamsters has delighted even some who never had a connection with the union, notably a dump-picker by the name of Emerson White. He can be seen on most winter weekends ice fishing on Finger Lake outside Wasilla. He wears $150 bunny boots, classic thermal coveralls, and expensive sealskin mittens — all retrieved at the Wasilla dump. "When those union guys headed for Texas or Oklahoma, I had it made," White says. "They didn't take back anything warm. They just drove up to the dump and threw stuff in by the carload. I guess it gets too warm down there to wear this stuff. Or maybe they just didn't want any reminders of Alaska."

CHAPTER 13
Eskimo Capitalists

NATIVES OF ALASKA have had to adapt in many ways over the centuries to survive the harsh extremes of the Great Land. For at least ten thousand years, Eskimos in the north hunted and fished the gale-whipped Arctic coast. Athabaskan tribes roamed the forested interior and the southeast, following caribou and salmon migrations. And the Aleuts on the bleak Aleutian Peninsula subsisted for centuries in underground shelters, and endured slaughter and enslavement by the Russians. But nothing changed Natives' lives as swiftly or traumatically as the arrival of the oil age in Alaska.

Almost overnight, the Alaska Native Claims Settlement Act of 1971 transformed Natives into capitalists. In exchange for surrendering their claims to ancestral lands, the settlement gave the state's estimated seventy-five thousand Eskimos, Indians, and Aleuts $962 million in cash and 44 million acres, about four times what all the dispossessed Indian tribes in the lower forty-eight ever won from the Indian Claims Commission.

Recognizing its past mistake in having herded Native Americans onto reservations, Congress instead devised a unique social experiment. Alaska was divided into thirteen regions, including one designated as "at large" for those Natives who weren't affiliated with a village. Each region was empowered to establish a corporation to manage its new wealth and develop its land. Anyone with at least a quarter Alaska Native blood—a minimum of one Native grandparent—became a stockholder and received one hundred shares in the regional corporation. Forty-five percent of the claims money was assigned to the villages in each region. The villages would also be permitted to organize into corporations, though their plans for the claim money and land selections would be subject to approval by their regional corporation. In short, Alaska Natives would be allowed to control their own destinies, but within a legal structure devised by white men.

Eskimos and Indians, who before statehood had been barred from many saloons and forced to sit in separate sections in some theaters, suddenly became Alaska's most sought-after citizens. As hundreds of millions in reparation dollars started flowing into corporate treasuries, Natives were pursued by hordes of developers, land speculators and securities salesmen offering investment schemes, and a mob of consultants, bankers, and lawyers offering advice—for a retainer—on how to guard against those who would fleece them. "I can recall stock salesmen inviting Native leaders to lunch, plying them with booze, and making glowing promises of high return on investments, which, of course, carried high commissions," Ralph Papetti, a veteran stockbroker in Anchorage, says. "The Natives were also besieged by people selling them trucks and small airplanes. Transportation was very important in the bush. After years of using dogsleds, powered vehicles had great status."

Many Alaskans, resentful of the hefty settlement for a frontier they believed whites had largely tamed, predicted the Natives would be parted from their money and land in a matter of years. But few would foresee the social impact of their being catapulted into the twentieth century. Anthropologists call the transition "rapid acculturation." Since the early part of the century, hellfire-preaching missionaries, BIA school disciplinarians, and patronizing social agencies had steadily imposed alien values on Natives and stripped them of much of their heritage. The money generated by oil heaped on new ideas and expectations. Natives eagerly bought television sets and satellite dishes with their first land claims dividends, seeing on their screens a good life that failed to materialize.

Men who had once provided for their families from what nature offered were now expected to look for cash jobs to pay for modern needs and conveniences. But except for the few village corporations that found oil on their property or could tax oil-producing facilities, the new order did not provide jobs for the masses. Many Native men moved into the cities to look for work, but few succeeded in getting into the mainstream. The sidewalks along Fourth Avenue in downtown Anchorage and the little park outside the old city hall are lined with disheveled men looking for a handout or sleeping off a drunk.

Back in the villages, women and their children can get along on welfare. The elderly collect Social Security. The modern market economy and government paternalism have robbed males of their self-esteem. They no longer are critical to family survival. For solace, and perhaps retribution, many men (and many women, too) have turned to comforts some of them learned working the pipeline. Alcoholism and drug addiction have given rise to hundreds of shelters, detox facilities, and abstinence programs. Troubled village leaders have made use of Local Option No. 4, which allows communities to drastically restrict or outright ban the use or importation of alcohol. The result has been significant population shifts, especially an influx of determined Native drinkers to the cities, as police and community service patrol squads can testify.

Particularly tragic is the record rate of suicide among young Native males who succumb to despair and the disorientation of drunkenness. Sporadic epidemics of suicide among teenagers are common in the villages. According to the most recent figures available, Native males between the ages of eighteen and twenty-four are killing themselves off at ten to fifteen times the rate of their peers in the lower forty-eight. Accidental death, in which alcohol is often a factor, occurs at five times the national rate. The suicide rate of young Native women is about five times the national average. But their dependence on alcohol raises other specters. Rates of both fetal alcohol syndrome and sudden infant death are more than twice as high as they are for the rest of the nation, and the consequences are costing the state millions.

Ann Walker, executive director of the Alaska Native Health Board, attributes much of the Natives' substance abuse to their sense of impotence in their changed roles. Uneasiness about their ability to succeed in the white man's world was, in fact, evident among Native leaders from the outset of their venture into capitalism. "We lived for centuries on the land and on what the land could provide, and all of a sudden we are businessmen," Nelson N. Angapak, the son of an Eskimo reindeer herder and a director of Calista Native Corporation, recalls. "Now the Natives are supposed to leave the subsistence life and compete with Wall Street."

Compounding Natives' problems, exploitation of their newly rich corporations became a world sport. Angapak's Calista represents 13,308 shareholders in the second most populous region after the one represented by Sealaska. Its largely Yupik Eskimo population inhabits the soggy, almost treeless coastal and delta plains where the Yukon and Kuskokwim rivers flow into the Bering Sea in southwestern Alaska, embracing a territory the size of Michigan. The region, home of the Yukon Delta Wildlife Refuge about 500 miles west of Anchorage, is widely known for its large seabird rookeries and varieties of migrating waterfowl. (It is the home and nesting area for all of the Cackling Canada and Pacific white-fronted geese in the world.) Less known is that it is also the home of the most economically and socially distressed people in Alaska, and possibly in the U.S.A.

Forty-eight year-round villages, most of them clumps of crudely boarded tar-paper shacks, are scattered throughout the region. They are widely separated, with no roads connecting them to one another, much less to the sparse road system of the state of Alaska. The people live on fish and waterfowl and whatever land mammals venture close to the Bering Sea. Little village stores sell nonperishable items to supplement subsistence harvests, at prices about 50 percent higher than in Anchorage. Except in Bethel, the Kuskokwim delta city where forty-six hundred people live, there are no hospitals or libraries in the region. In short, the region looks much as it did in precorporation days.

When news spread that Calista would receive $80 million in cash and 6.5 million acres of surface land, there was rejoicing. Unemployment in most villages exceeded 60 percent. No one could accurately estimate per capita income, but it was fair to say that most Natives in the Calista region earned less than $2,000 a year. The region's health problems were the worst in the state, with regular outbreaks of hepatitis, meningitis, and tuberculosis. The Kuskokwim area also had the state's highest levels of violent crime, suicide, and accidental death. Finally, it seemed, money was in sight to address these problems.

In fact, a lot of money did come into the corporation. But most of it went out just as quickly, without touching the lives of Alaska's neediest citizens in any meaningful way.

Calista's first problem was one common to all the Native corporations. Having won the passage of the Native claims act, Native political leaders naturally took control of the newly formed corporations. But they were not businessmen, nor did they possess the skills or experience to select trustworthy, capable professional advisers. Their real expertise was in tribal politics, so it is hardly surprising that many of their business decisions were political decisions. The old traditional ways did not foster prudent financial practices, and they resulted in liberal amounts of patronage but few profits.

"We will pursue every available employment for shareholders," Calista stated in its 1974 report. But when it tried to form a construction company that would provide jobs for its people in the Anchorage building boom, all it got was a painful lesson and a lot of red ink. "We bought a lot of equipment and incurred a lot of debt service," explains Mike Niemeyer, who is part Yupik and a vice-president of the corporation. "Later, we found it would have been more efficient to lease equipment. We also hired a large permanent staff when we actually needed only a small core of permanent people. The costs started to pull us under." The company never returned a profit and was losing about $4 million a year when it went under in Alaska's real estate bust of the late 1980s.

Calista also ventured into the fish processing business. Many of its shareholders fish the king and chum salmon that migrate up the Kuskokwim and Yukon rivers. "We bought a couple of collector boats and raised the price [at which we] bought from our fishermen and sold to the Japanese," Terrence Reimer, chief operating officer of Calista, says. "We tried to create a better economy to help support the subsistence life-style. But the price of salmon dropped everywhere and we no longer could afford to sell salmon at huge losses to benefit only a segment of our shareholders."

Calista directors thought it would be a fine idea to train their people in hotel management, since tourism was certain to become a growing industry in Alaska. So they set out to build a palatial hotel that would be not only a training base but a symbol of Calista's economic presence in Alaska. The sixteen-story hotel was to be the centerpiece of an office-retail complex to be called Calista Square on the eastern edge of downtown Anchorage. The wife of an early Calista president was chosen to be one of the interior designers. While other prestige hotels in Anchorage highlight their lobbies with a trophy-size stuffed grizzly or polar bear, she ordered a $700,000 solid jade staircase. It turned out to be too slippery to walk on, however, so it had to be covered with carpeting.

Cost overruns put the hotel in deep financial trouble before it was half finished and brought construction to a standstill. The Native leaders went to Washington for help. They returned with a promise that the BIA would provide a 90 percent guarantee on a $34 million loan—said to be the largest in BIA history—on the condition that Calista sign up a major chain to operate the finished hotel.

After many inquiries, Sheraton emerged as the most likely partner. A committee went to Boston to iron out the contract. They were wined and dined, in the words of one Calista insider, and came back with a deal he describes as "no-win for Calista and no-lose for Sheraton." Sheraton agreed to operate the hotel and train twelve Eskimos in hotel management. However, it insisted on a management fee based on a percentage of the gross revenue, assuring it a comfortable return whether or not the hotel was profitable. Further, the Natives were required to buy many of the basics from Sheraton, including towels, sheets, and napkins bearing the Sheraton logo.

A handsome hotel was erected, but the arrangement made it impossible to return a profit. After five years, Calista could no longer afford to underwrite the losses. It sold the hotel in 1988 to a Korean syndicate at a total loss of $40 million, and with little to show in the way of training for its people. "It was not as successful as I had hoped," concedes Martin B. Moore of Emmonak, one of the Eskimo directors who negotiated the deal in Boston. "Yupiks did not like the confinement of the hotel business. Some couldn't pass the test. Others couldn't stand the life-style. It is in the Native people's blood to return home when the weather is good. We want to go back where the air is free and the country is wide open."

As settlement payments continued to pour in, Calista tried new ventures but could not shake its management problems. It lost more than $6 million on an ill-fated surimi analog (artificial crabmeat) processing plant that it built in Olympia, Washington, with a Korean firm as a minor partner. Mismanaged construction resulted in a ruinous overhead. Oscar Mayer acquired the plant at a bargain and runs it successfully today.

Spurred by speculation that the state expected to build a bridge across the Cook Inlet connecting Anchorage to the undeveloped Knik Arm, Calista began developing a six-thousand-homesite complex called Settlers Bay on a scenic knoll near where the bridge would lead. The bridge was never built and only a few homes were sold. When it could no longer cope with the financial drain, Calista unloaded Settlers Bay at a $12 million loss.

Calista lost an estimated $39,009,000 during the three-year period ending December 31, 1988. An audit of the corporation's fiscal status that year contained an ominous warning: "The factors . . . indicate that the Company may be unable to finance its operations or meet its obligations as they become due and, therefore, be unable to continue in existence." While responsibility for the bad investment decisions must rest with the Native board of directors, it usually dutifully followed the recommendations of its white advisers. (The board was so inexperienced that it got rid of one chief executive, who had performed so badly that he placed the very corporation in jeopardy, by leading him out of his office and changing the locks on all the doors. The board's lawyer later told him he was fired.)

If poor business practices and naive leadership dashed the expectations of Calista, other Native corporations encountered more sinister perils. Bering Straits Native Corporation encompasses the Bering seacoast and Norton Sound settlements north of the Calista region. While its winters are more severe and its sixteen villages are similarly isolated, Bering Straits is relatively better off than its neighbor down the coast, with far fewer people—only sixty-nine hundred stockholders. Many of them find jobs in its largest settlement, Nome, of gold rush fame. The city's thirty-five hundred residents today have a bustling community of government employment and service industries.

George Bell, a Nome Inupiat Eskimo, was an incorporator of Bering Straits and a loan officer of the only bank in town, the Alaska Miners and Merchants. He was chosen to accept the first disbursement under the claims settlement act, which he promptly deposited in his own bank. Alaska Miners and Merchants had just been bought by Alaska National Bank of the North, whose president was Frank Murkowski, later to become a U.S. senator. The Bank of the North elevated Bell to director of its department of native affairs, in anticipation of the steady flow of claims payments from the government. Bell offered himself as financial adviser to his colleagues on the Bering Straits board, a position that enabled him to direct the millions in settlement money into the bank's coffers.

With $70 million assigned over the next decade, the Bering Straits Native Corporation's account became the bank's largest. However, as the money started to come in, Bering Straits neglected to turn any of it over to the villages, as mandated by the settlement act. Instead, it commingled all the funds and went on a spending orgy, making a series of investments through the Bank of the North. The bank, as a trustee, might have been expected to advise Bering Straits' directors whenever it had adverse information concerning a potential investment. But as a loan officer would later testify, the bank felt no such obligation. The list of unwise investments rapidly grew, and included a now defunct concrete conduit manufacturing company called Life Systems, which was having trouble meeting its payments to the bank even as it was unloaded onto the Native corporation.

Bering Straits lost $1.7 million on Life Systems, but that was minor compared to its other ill-fated ventures. Many more millions went down the drain when the corporation bought a construction company and its aging equipment from a white owner. The Natives hired a Panamanian to run it. He was instrumental in landing the company a big Seward dock project, but it underbid the job by an astounding $26 million, which it then had to pay to the bonding company. In an effort to cash in on the pipeline building boom in the mid-1970s, the Native corporation bought a tire-recapping company and started building a hotel in Fairbanks. Tire customers vanished with the completion of the pipeline, and the heavy debt incurred by the doomed construction company crippled the corporation's finances so badly that it never had the money to finish the hotel.

Newspaper accounts of depositions filed in a later court case indicate that during Murkowski's tenure the Bank of the North earned more than $1 million in fees and interest on loans to Bering Straits. The corporation was a banker's ideal client. As the parent company, it had guaranteed all the debts made by its subsidiary companies. The bank reimbursed itself for the bad loans by simply dipping into the stream of settlement money.

By the early 1980s, the Bering Straits Native Corporation was battling to stave off bankruptcy. It might have wallowed quietly in its misfortune had not the people of Sitnasuak, the Nome village corporation, decided to make an issue of the bank's role in the rapid depletion of the regional corporation's bank account. In 1976, and again in 1977, Richard Atuk, Sitnasuak's general manager, had written requesting a status report on an estimated $2 million due the village. Each time the bank stonewalled. In October 1980, the village sued the Alaska National Bank of the North in superior court, charging that the bank, as a trustee of Alaska Native fund monies, had helped Bering Straits "to systematically dispose of the village funds for the benefit of everyone but the villages." Under a promise that it would repay the money due Sitnasuak, the Bering Straits Native Corporation escaped becoming a defendant, and, in fact, joined the village in suing the bank until it ran out of money a year later.

The village continued to press the suit for five years before it came to trial in 1987. Both sides brought in high-powered lawyers—Charles Cole, today Alaska's attorney general, represented the bank, and R. Collin Middleton, an experienced trial lawyer from New England, pleaded for Sitnasuak. The village won a devastating victory. "The defendant Alaska National Bank of the North as trustee breached the obligations imposed by the trust causing the loss to the beneficiary, Sitnasuak," Judge Mark C. Rowland ruled. He ordered the Bank of the North to pay the village, which had a population of twenty-six hundred at the time, $14 million in damages. The bank went into receivership soon after, and the responsibility for paying off the court award fell upon the FDIC—a shifting of the burden onto the public that has become all too familiar.

Frank Murkowski, by now securely seated in the Senate, never personally appeared as a witness in the case. Middleton was obliged to go to Washington, D.C., and seek a dispensation from the Senate to depose him. The thrust of the senator's deposition was that he had been aware that Bering Straits was spending and encumbering far more funds than it had. He remembered that he'd once even called a meeting to discuss it.

Bering Straits Native Corporation filed for the protection of Chapter 11 bankruptcy in 1986. A number of other Native corporations, notably Calista, were also teetering on the brink of bankruptcy. Concerns rose that Alaskan Natives would, indeed, be parted from their money, and eventually their land.

Out of those fears emerged one of the most unusual special-interest tax giveaways in U.S. history. The right to sell net operating losses (NOLs) for cash to any profitable U.S. companies desiring to lower their tax bills had been the law briefly for all U.S. corporations in the early 1980s. But the practice created such a tax loophole that Congress made it illegal in 1984. Lobbied by the distressed Native corporations, Alaska's senior U.S. senator, Republican Ted Stevens, managed to persuade his colleagues in Congress to pass a special tax law in 1986 that reopened the privilege only for Alaska Native corporations.

The tax break saved several of the Native corporations from extinction. Bering Straits sold more than $250 million of hard cash losses, including one $55 million sale to Del Webb Corporation, which had been rolling in profits from building Sun City in Arizona as well as Nevada gambling casinos. Those sales returned about $80 million in cash, giving Bering Straits the resources to pull out of Chapter 11 and even to settle its debt with Sitnasuak. Calista Native Corporation's heavy losses from the ill-fated Sheraton Hotel and Settlers Bay real estate ventures enabled it to sell $70 million in hard cash losses, for a return of about $33 million in cash to its depleted treasury.

This was the way the tax break was supposed to work. However, lawyers for other Native corporations decided to get creative. They persuaded Native directors that NOLs did not necessarily have to be in hard cash but could also be land devalued when oil or gas exploration proved unproductive. They contended that the falling prices of timber and oil made the land less valuable than at the time of conveyance by Congress, constituting a loss accrued to the corporations.

The Cook Inlet Region Inc. (CIRI), by far the most profitable and one of the best-managed of the thirteen Native corporations, promptly multiplied its wealth by testing the limits of the tax loophole. The corporation reaped $102 million from losses created from devalued oil and gas reserves, almost as much as its poor cousins, Bering Straits and Calista, collected on their total hard cash losses.

CIRI, the Native corporation serving the Anchorage region, had much experience in getting the most out of Uncle Sam. Its original share of the claims settlement was a modest $78 million in cash and 2.3 million acres of land. It selected promising oil and gas properties in the Cook Inlet area but soon ran out of available land. It could pick up only about half of its allotment because much of the land around Anchorage was already in private hands or claimed by state and federal agencies. In exchange for the unavailable acreage, CIRI persuaded Congress in 1981 to give it $200 million worth of credits to bid at auctions of surplus federal property. Newspapers tracking its progress estimate that by 1991 CIRI built up a billion-dollar real estate empire that included more than thirty former federal properties, ranging from beachfront acreage in Miami Beach and Potomac River frontage in Alexandria, Virginia, to extensive holdings in Hawaii and California. CIRI also took advantage of tax breaks offered to broadcasting interests that sell their businesses to minorities. It now owns major interests in television stations in New Haven, Connecticut, and Nashville, Tennessee, as well as radio stations in Boston, Washington, D.C., Chicago, Seattle, and three other metropolitan areas.

CIRI's president is Roy M. Hundorf, born of a Yupik mother and a German father. Dressed in a well-tailored gray business suit, he parks his slender six-foot frame behind a large desk on the executive floor of CIRI's gleaming Anchorage high rise, which is tastefully decorated with Native art and artifacts. Occasionally glancing out the window at the Cook Inlet, he sums up the corporation's formula: "Our policy is to run a successful business operation," he says. "To the extent that we can, we hire shareholders, but we don't overburden ourselves. If we fail in business we are worthless."

CIRI earned $237,953,000 through 1988, compared to the combined $16,506,000 earned by the twelve other regional corporations since Native claims settlement payouts started twenty years ago. CIRI was luckier than the others in that it was the only one to find substantial amounts of gas and oil on its property. It had the further advantage of the $200 million in federal surplus credits, which it invested wisely. And CIRI comprises only seven villages, which made it easier to get a consensus on its policy to seek good investments first and jobs second. No one doubts that CIRI is going to survive.

The Kotzebue region's NANA Regional Corporation, headed by Willie Hensley, is also posting an impressive record. Hensley, a tireless Native activist still, stressed creating jobs and preserving culture from the time the corporation was formed. He won wide acclaim for developing the Red Dog zinc mine, which employs three hundred of his region's people.

But economists who have analyzed the twenty-year record of Native capitalism find that the majority of the regional corporations are in real trouble. Besides bad investments, much of their money has been dissipated on legal fees, often incurred by fighting one another. Endless lawsuits over sharing of revenues derived from the sale of natural resources, over grievances filed by dissident stockholders, and over challenges to subsistence rights have enriched Alaska's lawyers. One corporation officer estimates that lawyers have skimmed off at least $100 million in fees from the original $962 million claims settlement.

The special privilege of selling NOLs for cash may have stemmed the demise of some of the corporations, but it can be blamed for accelerating the self-liquidation of others. And it is the root cause of an environmental massacre that is ruining much of scenic Alaska today. In pursuit of revenues from NOLs, many Native corporations denuded Alaska forests and sold the timber—principally to the Japanese—at bargain-basement prices so that they could profit by selling the red ink from their huge losses. Environmentalists have long focused public attention on hefty U.S. subsidies that benefited two big pulp mills logging in the Tongass National Forest, ignoring this furious devastation of much of Alaska's original-growth rain forest, even as they wage hard-sell fund-raising campaigns to save the rain forests in South America, Africa, Malaysia, and Indonesia.

A visitor flying over Prince William Sound sees thousands of acres of clear-cut patches where ancient hemlock and Sitka spruce once created a wilderness home for deer, bear, river otter, and nesting marbled murrelets. Farther south, in Yakutat, fishermen pass miles and miles of devastated rain forest on their way to the Situk River, famed for its run of steelhead trout. This stately forest, which contains three-hundred-year-old Sitka spruce, was leased for logging by the local Tatitlek village corporation. It was supposed to be a sustained cut of twenty years with about 10 million board feet to be harvested a year, but crews wielding 36-inch Stihl chain saws have taken down as much as 60 million board feet a year since cutting started seven years ago. Today there is little left to cut. "Every agreement that the timber industry ever made with the community has been broken," says Larry Powell, the mayor of Yakutat. "Stream protection. Annual allowable harvest. Size of clear cut. Providing a sawmill to employ twenty to thirty people." (A sawmill was built during the first year but its roof caved in under heavy snow the first winter and not a single board was cut. Instead, the logs go directly to Japan, Korea, or Taiwan.)

The greatest desecration, however, occurs in the vast Tongass National Forest in southeast Alaska, the embattled 16.9-million-acre preserve of moss-draped trees, mountains, glaciers, salmon streams, marshy estuaries, and habitats for grizzly and black bears, Sitka deer, and bald eagles. Loggers from the two beleaguered Outsider-owned pulp mills in the area, whose practices and subsidies have been reined in by wilderness legislation enacted late in 1990, are pleased to steer visitors to Native corporation clear cuts in the Tongass that go right down to pristine salmon streams—a practice for which the pulp companies would have been quickly hauled into court.

One of the boldest Native corporations cutting in the Tongass is Shee Atika, headquartered in Sitka. On the verge of bankruptcy only a few years ago, it has been hurriedly marketing its 23,000 acres of prime timber on Admiralty Island, one of the unprotected wilderness areas. In trying to regain its fiscal health, Shee Atika pushed the 1986 tax loophole to the breaking point: first, tax breaks for hard cash losses, next for devalued land, then for timber loss sales, and finally for the $10 million sale in 1981 of timber appraised at $175 million to a new business named Atikon Forest Products—which is 49 percent owned by Shee Atika. After an adjustment for some harvested timber, Shee Atika claimed a $152 million loss, which it sold to Quaker Oats Company for $57 million. Meanwhile, Atikon—and by extension Shee Atika—stood to profit from processing the undervalued timber.

The IRS audited Shee Atika's loss claim and hit it with a $60 million bill for back taxes in 1991. If upheld on appeal, it could drive the Native corporation into bankruptcy. Some regard it as the government's opening salvo in an effort to recover some of the hundreds of millions of dollars in tax write-offs obtained through questionable tax manipulating by Alaska Native corporations.

Congress did finally close the NOL loophole completely in 1988. But the damage it caused is far from ended. Having given their shareholders a taste of hefty dividends, the directors of the Native corporations, who depend upon shareholder votes to stay in office, are under pressure to maintain them. The only way to do so is to keep cutting until the forests are gone, which some Native leaders concede will happen soon.

In 1990, the incorporated Tlingit village of Klukwan, which once advertised tax losses for sale in *The Wall Street Journal*, declared a dividend of $36,000 for each of its 253 shareholders. The windfall came from clear-cutting majestic, old-growth Sitka spruce on Long Island in the Tongass National Forest, which it obtained after it ran out of land to select in the Haines area, where the village is located. "The island has some of the finest timber in the world, and we knew we were going to become timber barons," Irene Rowan, chairman of Klukwan at the time, says. "The Forest Service approved it. There were no environmentalists around. So the state signed it off to us in 1977. Now we are being pressured for easements, the environmentalists are on us, and the government is coming back at us through the Internal Revenue Service."

"Don't you worry about the destruction to the environment?" she was asked.

"Let me explain from an Indian standpoint," Mrs. Rowan replies. "Our land was taken away from us when they declared Tongass a national wilderness. We spent half of our lives trying to get the land back. Now we are spending the rest of our lives trying to hang on to it." Pressed on the subject of exploitation of tax loopholes by Native corporations, she adds, "Wouldn't you try to make as much money as you can now? Our people feel that it is only a matter of time before they again are going to take the land away from us."

Once the Native corporations run out of timber to cut, some of them will also run out of cash. What happens then? Amendments to the claims act protect the land from liquidation sales. But such protections have never been put to a legal test. With two of the largest timber-cutting Native corporations already in bankruptcy protection, the right to collect debts by attaching Native land looms as the next big battleground in Alaska. In 1993, restrictions against selling stock in the corporations are due to end. This will certainly invite another threat to Native ownership—hostile takeover attempts, which until now have been barred.

The Native Claims Settlement Act was supposed to provide security for present and future generations of Alaska's Natives. Corporations were expected to earn profits, create jobs, improve living conditions, and protect subsistence resources. Instead, half of them enter their third decade uncertain of their future. Bering Straits and Chugach struggle to recover from Chapter 11 bankruptcy. Four others—the Aleut, Calista, Bristol Bay, and Koniag corporations—are skeleton operations, scarred by business setbacks and now simply trying to stay afloat. The other six, led by the remarkable CIRI, have established solid fiscal foundations, but they also will face harder times as natural resources give out and the pressure to maintain shareholder dividends remains high. (The thirteenth corporation owns no land, and its assets have been distributed.) The NOL loophole provided an infusion of $445 million on top of the $962 million appropriated by Congress. Now, Alaska's Natives have no further payments and no further federal tax breaks to bail them out. Meanwhile, the state refuses constant offers from cash-drained Native corporations to sell back scenic holdings.

As Natives ponder their predicament, the chain saws keep tearing down Alaska's vanishing rain forests. Much of the irreplaceable beauty of this Great Land leaves on barges piled with 40-foot logs, heading west across the Pacific.

CHAPTER 14

Here Is Free

WHEN SUMMER COMES to Alaska, with it come great numbers of salmon on their journey from the Pacific Ocean into freshwater streams and lakes where they spawn and die. The migrating salmon sustain the river ecosystems, perpetuating a chain of life in which Dolly Varden trout wait in the eddies to devour salmon eggs, grizzly bears prowl the banks to feast on the weakening fish, and ravens swoop down to scavenge the carcasses.

This strange rite of passage has gone on for more centuries than anyone knows. The other animals who depend on it divide the spoils, whether the year is abundant or lean. Only humans battle with increasing intensity over who gets what share of the harvest. Alaska licenses about thirty-three thousand commercial fishing boats and bestows on about thirteen thousand of them the ultimate privilege—a limited-entry fishing permit. In 1990, this exclusive group alone hauled in 690 million pounds of the five species of salmon from Alaska's designated waters on designated fishing days.

That is the legal harvest. However, huge ocean trawlers from the West Coast and the Pacific Rim, many with vacuum-cleaning efficiency, intercept untold millions of pounds of salmon illegally. They respect neither designated fishing areas nor other restrictions designed to protect the species. The trawlers deliberately poach the young fish in outer waters before they head for their spawning grounds in Alaska. International law forbids this, but the demands of the marketplace and the difficulties of policing the open seas make risk-taking profitable.

There are also Alaskan Natives who assert their subsistence rights and claim immunity from the white man's fishing laws. They net more than one million salmon a year, which are supposed to be dried and stored to feed their families. But many Native fishermen do not resist the temptation to turn salmon into cash when willing buyers appear. In a rare crackdown, L. George Schenk, a Bellingham, Washington, fish supplier, was arrested in the summer of 1992 for buying 53,000 pounds of king salmon from Natives who sold their subsistence fish. He was fined $50,000, sentenced to six months in jail, and his fishing boats were confiscated.

The discovery of oil in Alaska has put still more pressure on Alaska's salmon stocks. The oil firms brought in waves of affluent, recreation-minded employees. And the boom generated gigantic state revenues that made still more citizens affluent and recreation minded. This expanded Alaska population, which also includes a steady and sizable flow of military personnel, accounts for most of the 350,500 sport fishing licenses that Alaska now sells each year. These rod-and-reel fishermen demand a crack at the salmon, too, and their numbers give them political clout. They've shown their determination to reduce the lopsided harvests of commercial and subsistence netters by fighting in the courts, in the legislature, and sometimes on the banks of the salmon rivers.

In the fall, about 835,000 caribou, comprising twenty-five distinct herds, leave their summer feeding grounds in Alaska's lower half and head for their winter retreats around the state. This annual migration also has been occurring since before recorded time. It, too, has been a support system for animals and man, particularly for Natives who live off the land. Though involving fewer people, equally bitter vocal and legal battles now split Alaska not only over who gets to shoot caribou but also over who gets the permits to develop near caribou habitats. The state's constitution guarantees that the fish and animal populations are common property resources and therefore belong to all. But each year more parties lay claim to them while more people engage in shrinking the environment that supports them. It's a dilemma with a long and intractable history.

During the first half of the century, the territory sold out to the cannery interests, which nearly decimated the salmon by intercepting them with huge, deadly efficient traps stretched across the mouths of rivers. Most Alaskans cheered when fish traps were banned by the first act of the first state legislature in 1959. However, curbs on the canneries brought on a flood of independent fishermen. They saturated bays with gill nets and blockaded the migration routes near the shorelines with setnets—all legal equipment, but with the same devastating effect as fish traps when used in such profusion. By the mid-1970s, overfishing caused salmon stocks to crash again. From 1970 to 1975, the catch of Alaska salmon fell from seventy million to twenty million—and the number of licenses continues to grow.

State senator Clem Tillion, a strapping, bushy-haired salt who fishes off Halibut Cove, the picturesque harbor across from Homer in Kachemak Bay, was among the first to sound a public alarm. In the early 1970s, he saw that salmon stocks were approaching record lows in the once rich fisheries of Cook Inlet and Bristol Bay, while the number of fishing boats kept multiplying. He started a campaign to persuade Alaska to preserve its salmon runs by establishing a permit system limiting the number of boats and setnets eligible to harvest the fish commercially. Immediately, an uproar was heard around the state. "Up to then, fishermen had been setting their own rules, and many refused to accept any interference," Tillion says. "Kodiak in particular balked at any kind of regulation. They would sing the cowboy song 'Don't Fence Me In,' and tell everyone not to let that happen to Alaska." Tillion's telephone rang constantly, sometimes with threats to his life. His children were harassed in school and someone managed to slash a niece's tires.

Tillion won. In 1972, voters approved an amendment to the state constitution rescinding a common resource section that barred the state from restricting the numbers of fishing boats. Then, in 1973, the legislature passed Tillion's limited-entry bill. Henceforth, Alaska would regulate the number of commercial fishing boats, establish zones they could fish in, and set the dates and hours when fishing would be permitted. The prevailing understanding was that any Alaskan who had fished commercially earlier was assured of a permit, and the influx of boats from Seattle and elsewhere would be curbed.

A logical method of issuing the permits might have been to award them on open bidding or at least to charge a fee based on an estimated commercial return. But that would have practically shut out Alaska's Natives. With the nation's conscience still tender over its treatment of dispossessed Native peoples, the state rashly decided to give away the fishing rights free of charge. A commission determined eligibility by using a point system, with priority given to those who had fished in areas where there was no alternative employment. This tilted the advantage to the Natives. Points were also given to those who had fished for the greatest number of years and to those who had fished the most in recent years. By the time this giant giveaway ran its course, the state had handed out thirteen thousand limited-entry permits, the exclusive license to fish commercially. The majority went to gill-netters, who drift their nets into schools of salmon on designated days. About two thousand permits went to setnetters, who trap salmon by stretching gill nets from shore and anchoring them at a specified distance from the beach. Purse seiners, who use two boats in tandem to encircle schools of fish, won about the same number of permits as the setnetters, and about eight hundred permits went to power trollers, who fish designated waters in vessels with baited fishing poles extended from their sides.

But the legislature, under pressure from the fishing lobby, placed no restrictions on resale of the permits and attached no provisions for returning permits to the state when the owner died or quit fishing. Not surprisingly, many rushed to cash in what they had received for nothing. Licenses were traded on the open market, creating a lively exchange that continues today. Purchasers currently bid as much as $300,000 for gill net permits and $15,000 for setnet licenses. Their collective value is estimated today at $1.2 billion, according to a recent survey by the House Research Agency of the Alaska legislature.

The attempt to help Natives fishing in areas of no alternate employment did not work as intended. "Some Natives never understood the limited-entry process and sold their permits for little or nothing," Tillion says. "They now work on a fishing boat, and the permit is probably owned by some doctor in Anchorage." He could add Seattle businessmen, North Slope oil workers, and university professors to the list of those who have acquired permits from their original owners. Those with means are today the principal holders of the right to catch and sell Alaska's precious seafood without returning any substantial royalty to the state.

While the number of commercial salmon permits remains constant, the competition to harvest other fish has not abated. Permits to catch other species are open to anyone, and in 1991 there were about twenty thousand applicants to harvest halibut, crab, pollock, and even sea cucumber. The permit renewal fees for Alaska fishing boats range from $50 to $750 a year, depending on the species and the area. Fishermen and processors in Alaska waters are supposed to pay a nominal tax on their catch, estimated at $1.7 billion a year, but cheating is rampant, the Internal Revenue Service says. The House Research Agency estimated that fishing-related industries paid $46.6 million in taxes and license fees in 1989, while Alaska spent more than $60 million on programs to help them. And that net loss does not include the environmental damage caused by fuel spills, discarded trash, or the torn netting that floats out to sea and smothers thousands of fur seals each year, as the National Marine Fisheries Service reports.

"We didn't adequately tax the fishing resource because in the glory of oil we gave up all tax revenues from other sources," Tillion admits today. "Yes, it's a robbery." And he feels that nothing much will change as long as the state has oil revenues to take care of 85 percent of its budget.

The sellout of Alaska's prized resources doesn't stop there. The state's generosity, coupled with widespread lack of respect for the environment, is causing big trouble on shore, where greedy real estate speculators, unethical fishing guides, and lawless bank fishermen also exploit Alaska's natural treasures. Collectively, they are trashing the waters and shores of North America's most famed salmon stream, the Kenai River in south-central Alaska.

The Kenai's glacier-fed, turquoise waters are renowned for record-size king salmon (the world record, a ninety-seven-pound, four-ounce chinook, is mounted on the wall of a Soldotna auto dealer's showroom). The river also has a generous run of the two other most widely sought species, the sockeye and silver salmon, as well as trophy rainbow trout. As the state's single largest recreational resource, the easily accessible Kenai draws fishermen from the world over. Its headwater is the breathtakingly beautiful Kenai Lake, which carries the runoff of the snow-topped Chugach Mountains, about 90 miles south of Anchorage. The river rushes for 50 miles through the aspen, birch, and hemlock forests of the Kenai Peninsula until it empties into the Cook Inlet. Its deep pools, swirling eddies, and white-water rapids are a fisherman's dream and an artist's delight.

The peninsula itself is the original land of the Kenaitze Indians and is rich with Alaskan lore and history. The harbor town of Kenai, where Russia and the United States once maintained military forts, has had mining and fishing booms, but by 1950 its population was down to 321 people. Most of the white residents were homesteaders, many of them beneficiaries of the Veterans Homestead Act, which was passed after World War II. They cleared the required 20 acres, staked a claim to an allowable 140 acres more, and lived in chosen isolation. The surviving Kenaitze, having been exploited by the Russian fur traders and largely disdained by their new American neighbors, also preferred to keep their distance. Few, if any, pinned down ownership of their land as homesteaders. It was not the way of their culture, and in any case it would not have occurred to them that lands they had lived on for centuries were not their own.

When the Richfield wildcatters struck Alaska's first commercially productive oil well at the Swanson River on the northwest edge of the peninsula in 1957, this peaceful land was suddenly overrun. The population shot up by 2,500 percent over the next thirty years. An invasion of noisy, hard-drinking boomers, gamblers, and other oil camp followers not only ended the solitude of the Kenai but also trampled whatever remained of the ancient Kenaitze culture. Loretta Breeden, a homesteader now in her sixties, had been a reporter for the *Cheechako* News, a Kenai weekly born in the boom. She recalls a night on the police beat: "There was a shooting, three knifings, and someone blew up a building where they were having labor trouble. And police were asked to look for a gal who took off with several thousand dollars after selling tickets at the local saloons for an evening with herself."

Kenai became the first oil capital of Alaska. Camps and trailer campgrounds cropped up overnight on the banks of the Kenai River. No one raised questions about sewage disposal, bank erosion, or wetlands protection. And that was just the beginning. A dirt road connecting the peninsula to the state road system had been opened in the early 1950s to help the homesteaders. The discovery of oil made it urgent that the state pave it all the way to Anchorage. That was a signal for real estate entrepreneurs to go into action.

The banks of the Kenai were targeted by hard-sell developers, who tempted struggling homesteaders with money they couldn't refuse. Most of the lots were at least a half acre, but many were cut into parcels barely big enough to accommodate a cabin, much less a cesspool or a septic tank. If a lot didn't have a view of the river, the owner would often cut down the trees on the banks to make one. Those who didn't have direct access to the river sometimes found a way to manufacture it. Campgrounds carved boat slips into the banks and rented them for the summer. And in a coup that would make any Florida developer envious, the Kenai Keys, a 120-lot riverside development with locked gates, actually got permission from the U.S. Army Corps of Engineers to gouge out three canals totaling more than a mile in length, creating scores of additional river-access lots that sell today for $40,000–$50,000 apiece.

Having a place on the Kenai became a sign of status in Anchorage, and with weekend cabins came weekend crowds. Most were content to fish, but some hauled down powerful jet boats. Soon, small flotillas roared up and down the world's most famous salmon river, leaving 8-foot-high rooster tails in their wake and pounding waves over the fragile banks. Swamped fishermen, canoeists, and homeowners watching their river frontage wash away eventually besieged the state to halt the abuse. The property owners claimed high-speed wakes eroded the banks so badly their cabins were in danger of falling into the river. A few did suggest that the siltation caused by erosion was smothering salmon spawning beds, but those concerns made little impact because everyone was still catching plenty of salmon.

Neil Johanssen, an environmentally sensitive California transplant, has been director of Alaska's state park system since 1980. His 3.2-million-acre domain is the largest in America—about one third of all the state parkland in the nation. Yet he found that by 1985 the state had lost control of all but 15 percent of the Kenai riverfront. Homesteading and land selection by Native corporations had transferred an estimated 66 percent of it to private hands, and the remaining shreds were owned by either local municipalities or the federal parks system.

Nevertheless, Johanssen drew up a bill and with the help of sports fishermen lobbied the legislature to create a special management area under which a joint board representing the state and the municipality of Kenai would govern the river. If government could no longer get control of the Kenai's troubled banks, at least now it could lay down rules for use of the water that runs through them. The board's most effective action so far has been to ban jet boats from the waters and place a maximum thirty-five-horsepower limit on the size of outboard engines. But the measure is not enough to stop the deterioration of Alaska's famed salmon river.

No one feels the inadequacy more than Dan France, sixty-five, a longtime game warden who lives in a small house on a remote bluff where he can see miles of the Kenai. A short, round, energetic man with a full head of crew-cut gray hair, France remembers hiking with his schoolteacher wife, Mary, 3½ miles across a tangle of half-burned, fallen hemlocks to stake a homestead claim on this spot thirty years ago. Instead of a road, they built an airstrip, and commuted to their jobs by Super Cub until they could retire to the tranquillity of the river. They had expected some problems with bears, but soon learned that bears will stay away as long as you don't leave around food or anything plastic they can chew. Pine grosbeaks peck at the sunflower seeds in the feeders outside the window, and juncos and crossbills flit about the ground, eating the spill. An unsightly pile of lumber near the feeder is a hiding place for ground squirrels. France sits at his picture window for hours at a time, training powerful binoculars on the river. Watching the Kenai flow has not turned out to be the pleasure he expected. "We sit here watching the death of the river," France says, and Mary nods in agreement.

France dedicated twenty-five years to trying to save the river. Appointed a game warden by President Harry Truman, France had been assigned to Idaho and came to Alaska in 1954 as federal game warden for the territory. When the state troopers created a fish and wildlife protection unit, France donned their uniform and made the river his principal beat. His years of patrolling the banks left him with only contempt for the way Alaskans treat the Kenai. When the salmon start running, poachers go into action. They cover the mouth of a tributary with a seine at night, then send someone upstream to pour chemicals—usually Clorox—into the water. The salmon, which are highly sensitive to changes in the purity of water, try to escape downstream, where they are trapped in the nets.

France arrested hundreds of poachers each year. "Do you think it did any good?" he asks. "Let me tell you what I see out of this window." The problem starts with licensed fishing guides, he explains. About four hundred powerboat and drift guides offer services along the river when the salmon start running. They promise big fish and lots of them, and for a while the promise was easy to keep. "Many of them come up the river to fish right below me," France says. "I see the boat catch its limit of kings. They rush down the river, drop off their fish, and come back with the same people for another load. How long do you think this can go on before the fish give out?"

The drop-off boats, some of them run by guiding services, are also culprits. "These boats pull up to the banks near a likely salmon hole," France continues. "First the adults unload the beer and the soda, then the kids and the dog get out. More boats pull up. The same thing. The banks are pounded down by hundreds before the day is over. Banks crumble into the water. The bushes become a toilet. How much people-pressure like that can the river stand?"

Yet this destruction is mild compared to what occurs on the banks accessible from the highway. The confluence of the Kenai and the Russian River upstream of the Sterling Highway, about 90 miles south of Anchorage, attracts hordes of fishermen who stand elbow to elbow thrashing the waters. The mob scene is aptly known as "combat fishing." Signs say "Fly Fishing Only" and the law is that anglers can keep only those fish hooked in the mouth. Not only is the return rule widely ignored, but when night falls, whole families armed with lethal three-pronged grappling hooks snag salmon by the dozens.

Visitors from out of state—as well as many from Europe and Asia—come to the Kenai to fish for salmon on the cheap. They jam the adjacent $6-a-night state campgrounds, fill freezers with fish, and send loads of salmon back home. They boast that what they'll get for those fish from friends and relatives will pay for the trip. I asked one German why he didn't go to Norway to fish for salmon, since it was much closer. "Too expensive," he replied. "Everything there is private. Here is public. Here is free."

By the end of the salmon season, the banks of the Kenai resemble a battlefield, strewn with coils of monofilament line, fish lure wrappers, and discarded lunch bags and beverage cans. Paths through the bushes are filled with toilet paper and human waste. The streamside vegetation has been trampled and thousands of boots have left their scars on the banks. The damage caused by wading fishermen and churning motorboats particularly concerns Johanssen. "Out-migrations of smolt [juvenile salmon] occurs along the edges of the river," he says. "When banks crumble, siltation covers the gravel and robs the young fish of shade protection. And silt can smother any eggs that are hatching."

In the summer of 1990, Alaskans began to understand that the Kenai was suffering badly from the pressure. The return of king salmon, the largest and most prized species, dropped so severely that the state declared a catch-and-release-only moratorium—over loud protests from guides who make $600 to $1,000 daily in the king fishing season. The next summer, the number of kings had declined even more, by 35 percent compared with the poor season the year before. The state declared an emergency and again ordered fishing by catch and release only. Johanssen cut the number of licensed guides to 250.

Tim Hiner, a former Michigan guide who has worked the Kenai for the past fourteen years, concedes the river is in trouble. "As Kenai's population grew, more people fished it, and the Kenai's reputation for holding the world's largest salmon spread," he says. "Chances of catching a fish over fifty pounds are better here than anywhere in the world." Much of the blame rests with the greed of developers, he says. "The six miles below the Soldotna bridge [the stretch where salmon first enter the river] is virtually a slum. Trailers are parked up and down the banks. Rocks are rolled in where natural barriers are worn down. The aesthetics of this part of the river is terrible. There is no consideration for nature. People come up here and say, 'I never thought it would be like this in Alaska.' "

Hiner has strongly complained to the Kenai Special Management Board about the proliferation of guides as well. While state records show it issued only 280 permits for powerboat and drift boat guides in 1990, the scene on the Kenai at the peak of the salmon season makes a mockery of the figure. Boats are so thick from Soldotna on down a person could practically walk the 6 miles to the inlet by stepping from boat to boat. Hundreds of these vessels are owned by moonlighting guides who are content to take the small risk of arrest for the chance to make a fast $1,000 over a weekend. It is common knowledge that most of the time state police assign only one safety officer to patrol the entire river. "Let's face it. What river in North America has or needs more than 100 power boat guides?" Hiner wrote to the management board. A few days after he sent the letter, Hiner noticed a new neighbor, a guide who had just moved up from New Mexico. He hung out his shingle the same day.

"Overfishing has decimated the great king salmon runs in the Kenai but don't lay all the blame on guides," Hiner says. "There are also too many setnetters in the inlet and too many salmon are killed in deep-sea nets." He says he is seeing more kings arrive with scars from intensified netting. He posts pictures on his bulletin board showing salmon with noses bloody and twisted and tails split apart, signs they escaped the commercial netters.

Captain Phil Gilson, the chief fish and wildlife enforcement officer for the state troopers, readily concedes that the Kenai is underpatrolled, but feels the department must concentrate on catching the biggest offenders. "Is the state served best by having a man on the banks catching a couple dozen illegal snaggers, or by chasing commercial fishermen in closed waters where they might be taking three thousand salmon illegally?" he asks.

A fisherman himself, Gilson glumly notes that not only kings but even the silver salmon are not returning to the river in accustomed numbers. He blames the increasing perils that the fish confront as they try to reach their traditional spawning grounds up the Kenai. Young salmon start their journey to the sea after a year in fresh water. They depend upon shady, steeply banked corridors to help hide them from their natural predators, including Dolly Varden trout and mergansers and other seabirds. As the banks are worn down by people, their protective cover diminishes.

Things are no better once the survivors reach the ocean. "The usual stay at sea, which is four years, lately has been fraught with new danger," Gilson explains. "Japanese, Korean, and Taiwanese high-seas net drifters are fishing closer to Alaska. They let out miles of monofilament net forty to sixty feet deep. Supposedly they are fishing for squid. In truth, they catch anything that swims, including many salmon." Alaskan newspapers regularly report that American pollock fishermen, who operate on the high seas with giant sock trollers, are also becoming a high-seas menace as the market for bottomfish grows. Many king salmon, a protected fish, are jammed into the nets by the fast-moving boats, and only those caught at the very end survive. Since it is illegal to keep them, pollock boats dumped sixty thousand dead kings overboard in 1990, Gilson said, citing National Marine Fisheries Service figures.

Once the salmon make the turn into the Cook Inlet, they have to escape the licensed gill-netters. Hundreds of boats, each permitted to launch 900 feet of net, jockey for position as airplane spotters steer fishermen to the schools. Competing fishing crews have been known to level guns at one another.

Next, there are the setnet fishermen. Gilson says he doesn't think there is a vacant spot on the beach the entire 40 miles of the inlet during open season.

As the fish enter the mouth of the Kenai, the sports fishermen finally get a crack at them. The guides, who have made a science of salmon fishing, know how fast the fish travel, when the fish rest, and when they move. They can follow the schools all the way up to the confluence of the Kenai and the Russian River, where they stop—out of concern for their customers' safety amid the barrage of barbed, flying steel from shore.

"We will have to limit the number of users," Gilson says. "Catch and release is a possible remedy. Using only artificial bait could cut down the catch. I think we are seeing that the kings can no longer take the pressure." But Dan France is more pessimistic. "Forget the kings; they're gone," he says bluntly. "Concentrate on saving what's left of the river."

The intensity of the conflict over fishing is matched by the struggle over hunting that continues to polarize the state. The public perception was that under the Native Claims Settlement Act of 1971,

Eskimos, Indians, and Aleuts relinquished claims to their ancestral lands in exchange for 44 million acres of land and $962 million. But most Natives never dreamed this would affect their right to hunt, fish, and gather plants in traditional style. Meanwhile, environmentalists didn't foresee the destruction of natural beauty that would occur when Natives started cutting old-growth forests and otherwise transforming their resources into income-producing assets.

In an attempt to appease both groups, Congress in 1980 passed the Alaska National Interest Lands Conservation Act (ANILCA), which set aside 104 million acres as national parkland and wildlife refuges. Hunting is generally restricted in such areas, but the federal law also mandated that Alaska allow subsistence hunting by rural residents and fishing preference for those residents on all federal lands. The act effectively wiped out millions of acres of prime hunting territory for all but subsistence hunters. Tensions mounted further when some Native corporations closed their land to hunting except by their own people. Other Native groups charged hunters a fee to cross their lands, and at least one village now posts a specific kill fee for non-Native hunters: $3,000 per bear, $2,000 for moose, and $1,000 for caribou.

Alaska hunting guides, who virtually guarantee a trophy with any $10,000 wilderness hunting trip, might have been expected to protest. But the fraternity knew when it was well off. In 1974, the state had established exclusive guiding areas, which it gave away free to favored guides in exchange for a pledge to protect the stock of big game on their domain. A politically appointed guide board handed out exclusive areas to about two hundred guides of their choice, causing scores of others who were shut out to complain that criteria were based more on political connections than outdoorsmanship. Further, the state passed a law making it a felony to guide without a license. (Illegally practicing law or medicine is only a misdemeanor in Alaska.) With airplanes helping them spot bear, moose, caribou, and wolves, guides with exclusive territories could gross up to $500,000 a year. Craig Medred, outdoor editor of the *Anchorage Daily News,* reported in an excellent series on guiding that it was not unusual for guides to make a further killing by transferring their areas to others for prices as high as $250,000, even though the state still owned the land.

The injustice of the system angered Ken Owsichek, a native of Kenosha, Wisconsin. Concluding he could not guide in Alaska on his talents alone, he sued the state, charging that it had conspired to create a hunting guide monopoly. Meanwhile, a hunter and angler named Sam McDowell stepped forth to challenge the law giving rural residents priority in hunting and fishing privileges. A Missouri farm boy who came to Alaska in 1948 with $6 in his pocket, McDowell lined up a Native and an urban resident as coplaintiffs and went to court. Hunters and fishermen, many seething at the privileges granted Natives, helped McDowell raise more than $200,000 to fight the case to the state's highest court.

In 1989, the Alaska Supreme Court found the state guide law unconstitutional, ruling that it restricted opportunities to kill wildlife in the state. The next year, the same court ruled that the state subsistence law was unconstitutional as well, in that it illegally discriminated against city residents, notwithstanding the federal government's guarantee of subsistence rights for rural residents on its lands. Three years later, Alaska officials are still dancing around the tender question of Native rights. To resolve the issue would require either amending the state constitution to comply with the intent of the federal ANILCA act or suing the federal government on the remote chance that the federal subsistence privilege might be ruled unconstitutional.

Two consecutive state legislatures, in 1990 and 1991, could not muster the political courage to address the subsistence dilemma. When a third legislature tried to duck the issue in 1992, Governor Hickel convened a special session and ordered the legislators not to adjourn until they passed a solution. He had to post state police at the Juneau Airport to stop legislators from sneaking out. But even that drastic measure failed. The legislators passed a meaningless bill that attempted to define a subsistence user but fell far short of addressing the Native rights issue.

The chaos has thrown all hunting laws and some fishing regulations into disarray. Tensions between Natives and urban sportsmen are boiling at the highest level since the debates over the Native claims settlements in the early 1970s. Subsistence hunters who used to get first crack at the Nelchina caribou herd have been made to wait until the season opens for everyone. Three Natives in the village of Togiak were arrested for forcing white fishermen at gunpoint to leave a remote river. Newspapers regularly report on defiant Native hunters who kill migratory birds and caribou in their traditional way, regardless of season or quotas. The village of Akiak even officially decided to conduct its traditional caribou hunt in the summer of 1991, when villagers knew it would be illegal. Meanwhile, the insensitive Alaska Board of Game created a national storm in 1992 when it advocated reducing the Fairbanks area wolf population by authorizing a hunt from helicopters.

The polarization is enough to make Kreg Thometz, twenty-seven, an assistant hunting guide with Alaska Trophy Safaris for seven years, rethink his career. He fears that the controversy and the competition is putting big-game hunting in jeopardy. "Many guides are turning to fishing," he says. "I think I may too."

He promises to stay clear of the Kenai River.

CHAPTER 15
Power Politics

It is early May, and I am on my way to Juneau to watch the windup of the state legislative session. This is an exciting time in the state, but I can understand why so many dread the trip. Hemmed in by the Gulf of Alaska on one side and a mountain range on the other, Juneau is the most isolated state capital in the United States. All its roads eventually lead to dead ends. No highway or railroad connects it with the rest of the state or, for that matter, with the continent. The Alaska Airlines jet seems to be groping in the fog and rain as it approaches the airport, which sits on a wet plain amid Alaska's perpetually drenched panhandle. Even when the plane finally breaks through the overcast, there's no relief. Now it dips up and down on bumpy air currents as it passes through a narrow corridor, with mountains on both sides at eye level. It finally puts down, and you wonder why the capital of this wide-open state has such a tiny runway.

If Juneau's isolation daunts visitors, it couldn't better suit those who come here to make a deal. Legislative sessions run like small-town city council meetings. With only twenty members, the senate is actually a club. Eleven people can control it. Personal friendship rather than party affiliation, seniority, or ideology determine its presidency and committee chairmanships. On most days, both houses drone along in formal sessions, with rarely a spark of open debate. Proceedings are constantly interrupted by lawmakers asking for the privilege of the floor to introduce visiting constituents, including some who probably come into the gallery only to get out of the rain.

The real business of the legislature is done in the corridors, which are lined with lobbyists. Here you can see the selling of Alaska firsthand. Not only private interests but school districts, municipalities, Native corporations, and even nonprofit institutions pay big money to support a voice in Juneau. When the day's session ends, there is no place for legislators to hide. Lobbyists just walk a few blocks down the hill and continue to pursue commitments in the bar and dining room of the Baranof Hotel. They sometimes throw secret parties for favored legislators in the condos and the several mansions of this onetime gold mining city. But what transpires rarely remains confidential, because by the time votes are counted, almost everyone knows which legislator is aligned with which special interest.

Lobbyists nail down votes with campaign money or the promise of money. I was astounded to see legislators actually look to the gallery for a signal from their favored lobbyist on how to vote when a complex issue reached the floor. Even an image-conscious legislator like veteran senator Arliss Sturgulewski is philosophical about the lobbyists' power. "The legislature says, 'Here's the playing field, folks. Get in there. If you don't, you might be left out.' " Those who know how to play do very well. One of the most successful is rugged Sam Kito, a part-Tlingit Indian who in 1984 pleaded no contest to a charge of delivering cocaine. He earned $359,075 in 1989 from seventeen clients, largely because of his ability to get the ear of lawmakers who serve on key committees. Kent Dawson, a former chief of staff for Governor Jay Hammond, does very well too—$348,083 in 1989.

As unlikely as it seems, however, it took the oil industry a while to learn how to play the game. Not that the industry wanted political power for its own sake. When the last oil is pumped out of Alaska, the companies will move out. The industry wanted only to control the way it was taxed, to protect the cost of doing business from the vagaries of the political system as long as oil does flow.

For a while after the first commercially productive well was struck at Swanson River, the industry enjoyed a free ride. It paid no real estate taxes on its refineries, wells, or equipment staging areas. Firms and residences located outside the jurisdictions of the taxing cities were also exempt—there simply was no legal body to tax their property. Needless to say, many vested interests besides oil would have liked matters to stay that way.

After statehood came in 1959, the job of creating suburban municipalities with taxing authority fell to John Rader, a Stanford-educated lawyer then thirty-two years old, who was appointed attorney general by Alaska's first governor, William Egan. The few cities and boroughs that existed at the time of statehood had taxing power, but as the state developed, whole new settlements without taxing power grew up around them. No one outside the municipal boundaries paid any money for their children's education, although they sent them to schools financed by the municipalities. The citizens of Anchorage were particularly burdened because they paid to educate the children of Eagle River and Girdwood, two booming settlements just outside the city limits.

In 1962, Rader won a seat in the state house of representatives. He promptly pushed for enabling legislation to end the taxing inequities. Rader still wonders how he ever got the highly controversial bill through Juneau. Some of the press referred to compulsory boroughs, large or small, as "obnoxious." A prominent Republican legislative leader threatened, "Rader, I'll have to beat your head off with that bill if you push it." Antigovernment rumblings were heard across the state.

But the "ice bloc" legislators who represented the Eskimos from the lower Yukon had been having a problem with dog teams running loose. A child had been attacked and killed. Reindeer herders were complaining they could not protect their herds. And strays frequently got into vicious fights with tethered dogs. Rader persuaded them that a local government could license dogs and maintain a pound to confine problem animals. (He made his decisive pitch by taking off his clothes and cornering a powerful Native chief in a sauna.)

Yet Rader had no leverage with the powerful vested interests—the canneries, miners, and timber cutters. They had fought hard against statehood because they did not want to be taxed, and they were not about to change their minds now. Rader had assumed that the oil industry would be similarly hostile. He got the surprise of his life when he met with its lobbyist. The man listened attentively and then told him his clients would not object to paying their fair share of local taxes for schools, roads, police, and health services. They would object only to paying what they considered to be more than their fair share.

"I was shocked," Rader says. "They had to know what was coming [out of the Prudhoe Bay exploration] to decide to come out on the responsible side." The industry example proved crucial in breaking down the resistance of the interests that tried to hold out for a free ride.

Rader didn't encounter industry opposition until after the huge oil pool on the North Slope was discovered in 1968. Now a member of the senate, he led a coalition of Democrats and Republicans pushing for tighter controls on tanker safety. "Oil was now big and powerful, but it did not have a constituency at this point," Rader says. "They were ludicrous in political influence. They looked to the Republicans for help, but found that in Alaska few of the party leaders were doctrinaire." Governor Jay Hammond, a former hunting and fishing guide now in his second term, and Clem Tillion, a power in the senate and a professional fisherman, were Republicans, but they kept a distance from oil. While the industry fumbled for a presence in Juneau, the little senate coalition in the legislature pushed through the toughest regulations the industry would ever face in Alaska.

Rader's close colleague in the senate, Democrat Chancy Croft, a rangy, Texas-born attorney who also served a term as senate president, became an even tougher challenger to oil. (Croft grew up in Odessa hearing stories about oil company greed, and colleagues say he made it a mission to impart a portion of this distrust into legislation that would keep the industry in line in Alaska.) "When the oil industry changed from drilling on land at Swanson River to offshore exploration in the Cook Inlet, platforms were breaking down," Croft recalls. "But there were no state regulations, no state controls. We did manage early to pass a transport liability law, but that was not nearly enough."

One day Croft dropped into a bookstore in downtown Anchorage. "The clerk stops me and says, 'Here's something you ought to read, Chancy,'" Croft recalls. "She hands me the book *Supertanker!*" Croft read that oil was now being transported by container ships so large they could not make a quick turn and needed 6 miles to come to a stop. And these were the vessels that would soon be plying Alaska's tricky, environmentally pristine inland waterways.

By 1976, Croft pushed through a law designed to set up a $30 million fund for an immediate cleanup response, which he refers to as "sort of an oil spill fire department." The program was to be financed by premiums assessed against everyone transporting oil. Since premiums would be based on the safety of the ships—the less safe the ship, the greater the premium—Croft thought the oil industry would be encouraged to build safer vessels. Instead, the companies were angered by the state's attempt to inflict an added cost. Led by Chevron, the industry sued and managed to have the law declared unconstitutional two years later on the grounds that states could not mandate anything affecting the interstate transportation of oil. "I felt the attorney general was less than aggressive in defending the state's interest," Croft recalls. "In fact, an internal memo questioning the constitutionality within the attorney general's office found its way into the hands of the oil companies—a fact which came out in the trial."

Undaunted, Croft next succeeded in championing an income tax on oil based on separate accounting of the value of resources specifically taken from Alaska. Until now, profits from Alaska operations had been kept secret. The industry preferred the less revealing and less punishing apportioning tax formula that enabled international companies to hide Alaska profits by commingling them with profits and losses worldwide. Croft reasoned that since most of the oil flowed out of Alaska from one large oil field, profits were bound to be higher here than in states where oil had to be pumped from a series of smaller wells, resulting in less efficient operations. The figures the companies were forced to provide proved him right. Alaska legislated separate accounting, gaining about $1 billion a year in revenue from 1979 to 1981, though rising oil prices were partly responsible for the dramatic increase.

The industry promptly went to court again, trying to block the state from access to information on Alaska earnings. But the limits on state powers did not seem to be nearly as clear-cut this time. The case was still in litigation in 1981, but by then the Supreme Court had ruled on two cases involving separate accounting in other states. Neither was definitive for Alaska, but some of the court's language raised serious concerns in Governor Jay Hammond's mind, particularly a reference to separate accounting as "incommensurate" with apportioning tax. Meanwhile, the spending madness going on in Juneau compounded Hammond's fears. The Meekins coup of 1981 had divided the surplus oil royalties into thirds for the senate, the house, and the governor with the understanding there would be no challenges to how any branch spent its share. Intoxicated by the deal, legislators splurged the billions of oil money on pet capital projects as fast as they rolled in. A less selfish body might have held back some of the funds from the separate accounting windfall and placed them in escrow to protect the state in case it lost the court suit. But no legislator was about to give up his share of the oil booty.

Hammond calculated that by 1985, when a court decision could reasonably be expected, the funds in dispute would total more than $7 billion with interest. What if the state was forced to refund that amount? The oil industry used the specter of bankruptcy aggressively. Hammond decided not to gamble. Tom Williams, the state's commissioner of revenue, wrote the amending legislation that would repeal separate accounting. Senator Ed Dankworth, a former state police commissioner who was rapidly emerging as a friend to the oil industry, pushed it through the newly Republican-dominated legislature in 1981.

Only a few voices of protest gave Hammond pause when the bill reached his desk. Gregg Erickson, an economic researcher for the legislature, was one. He tells of trying to get the administration and legislature to view oil's relationship in perspective: "The oil industry takes five dollars out of Alaska for every one dollar it invests. They have one mission: to take oil out of the state as fast as possible and at the least expense possible. I questioned the reasoning behind this tax break and calculated it would cost Alaska billions. I believe that forecast turned out to be correct."

Hammond signed the bill. Nowhere in the language of the amendment was there any hint to the public that this was a gigantic tax concession to the oil industry. Separate accounting was killed in one succinct sentence: "Section AS43.21 is repealed." When the Alaska Supreme Court finally handed down its decision in 1985, it vindicated Croft's vision. The court upheld the validity of separate accounting. But by now it was too late. No one in the legislature rushed to restore it. Oil had learned how to play the game in Juneau.

Tom Williams, the revenue commissioner who had authored both the original separate-accounting tax bill and the revoking amendment, is today manager of British Petroleum's Alaska tax office. Despite his present affiliation, he thinks the legislature's failure to act after the court ruling ignored the premises under which Hammond had agreed to the amendment. "It was my understanding that Governor Hammond chose to go the safe route rather than gamble because he was certain the legislature would promptly reenact separate accounting if the state won the case," Williams says.

Chancy Croft, now absorbed in his private law practice, takes a grimmer view of the legislature's failure. "The demise of separate accounting marked the day that Alaska capitulated to the oil industry. That's when the state became a seller rather than a regulator of oil. Separate accounting not only gave us added revenue, but it also gave us important information. The state's power to tax is limited, but information can give you a good idea of what the limits are. It gave the state some basis to form an intelligent policy towards the oil industry, and Alaska lost that."

Alaska became not only a seller of oil but a cheerleader for the industry. Budget deliberations in Juneau now focused increasingly on the price of oil and the volume pumped into Valdez that the state would be able to tax. Newspapers headlined every half-dollar change in the price of a barrel of oil, every incident that affected the flow of oil, and how every such blip lowered or increased the next dividend from the Permanent Fund. The public got the idea that the higher the price of oil and the more barrels sold, the richer everyone became. And that became Alaska's bottom line.

The malady of greed that overcame the state in the 1980s assured the oil industry virtually complete protection from future regulations. Legislators were so numbed by the billions pouring in from Prudhoe—even under the industry's preferred tax accounting system—that interfering with tax formulas or tanker safety was far from legislative minds. The pervasive mood of the era is described by Senator Sturgulewski: "I can recall one legislator with a hundred million dollars at his disposal. He kept on saying, 'It's my money, it's my money to allocate where I want to,' as though he couldn't believe it. Could there be any doubt how this legislator would vote on any bill affecting the health and welfare of the goose that was suddenly laying those kinds of golden eggs for every lawmaker in Alaska?"

"No one calculated the price," Croft says.

"Everyone just counted the money," Sturgulewski adds.

As John Rader, Chancy Croft, Clem Tillion, and other influential reformers opted out of the legislature to pursue private careers, oil's influence in Juneau strengthened. The departing legislators had been household names, but a new breed of largely unknown candidates proliferated with Alaska's expanding population. They needed name recognition, and they courted oil's financial support to pay for the television and newspaper ads to attain it. "We lost people who had a historical perspective," Senator Sturgulewski recalls. "A good part of the time now was spent reinventing the wheel."

Once campaign money became that important, oil was in the driver's seat. For a time, Big Labor with its hefty war chest from pipeline construction had vied with Big Oil in Alaska election spending. Labor supported mostly Democrats and oil supported mostly Republicans. In 1986, each side contributed more than $400,000 to state candidates and parties. The financial problems and then the death of Teamster boss Jesse Carr set labor back soon after, but in any case the industry was already well on its way to persuading Alaskans that "They Are Us," a slogan it would come to use widely. Oil is Exhibit A of the fruits that flow to the public from developing Alaska. The industry found common ground with the miners, the timber interests, and all the other "Don't lock up the state" advocates. The new breed of candidate—Republican, Democrat, or independent—pounded on the theme of "more development." Most attached the now-cliché phrase "with proper environmental controls." Labor-backed candidates usually added the proviso of local hire.

Meanwhile, the oil industry was moving more of its people to Alaska and encouraging them to become active civically and politically. Many of them did, quickly becoming leaders in the chambers of commerce, citizen action groups, and political parties. Oil company employees became an important source of campaign money and voter clout. In deference to the realities, the voices of oil's traditional foes, the Democrats, were muted. Instead of questioning whether oil was paying enough, Democrats now just criticized government for spending too much. By the late 1980s, oil had everything under control.

The industry was quite comfortable with Alaska's all-Republican congressional delegation, which has been in place since 1981 (and even survived the Democratic national resurgence of 1992). Senior senator Ted Stevens, a lawyer, and Frank Murkowski, a banker, pushed oil's development agenda on the national front. Don Young, Alaska's lone House representative, who was often ridiculed in the press (he sometimes went elk hunting when Congress voted on important issues), faithfully carried out oil's errands. The presence of a Democrat, Steve Cowper, in the governor's office didn't bother the industry. Oil contributed generously to gubernatorial candidates of both parties, but it resisted the inclination to dominate the governor's office. Instead, it concentrated on maintaining control of the state senate, where throughout the entire 1980s it was able to block every threat to its bottom line.

What oil had finally learned was that sending smooth-talking lawyers to lobby the freewheeling legislators in Juneau was futile. Alaskans suspected Outsiders, particularly Outsiders in Brooks Brothers suits. The oil firms looked for an Alaskan voice and they found one in Bill Allen, the head of Veco International Inc., an oil services company that held lucrative contracts in the Prudhoe oil patch. Allen needed no persuading to champion the industry's cause in Alaska.

A high school dropout, Allen came to Alaska in 1968 as a welder working for a contractor building an ARCO-Alaska oil platform in the Cook Inlet. He branched out and developed Veco, a firm that could perform assorted other jobs needed in the oil fields. A bad investment in a Houston shipyard forced his promising company into Chapter 11 bankruptcy protection at the start of the 1980s, and he needed oil field contracts to get Veco going again. Allen did not have the skills or the prestige to sell the oil industry's message to the legislature, but he had an important friend who could—"Big Ed"

Dankworth, the former head of Alaska state troopers who had served in Juneau as a representative and as a senator. Dankworth had left the senate in 1982, the year he beat an indictment on two conflict-of-interest charges involving an old pipeline camp he tried to peddle to the state as a correctional institution. Allen promptly hired him as Veco's lobbyist, and "Big Ed" moved right back into the inner circle of the state senate as though he'd never left Juneau.

With Veco rising fast on the charts of the state's largest campaign contributors, Dankworth was welcomed back into the club. As a senator he'd known how to reorganize the senate leadership; now, as a lobbyist, he continued to do it behind the scenes. And he did it to the oil industry's liking, which is what Veco was paying him to do. Newspapers referred to Dankworth as "the oil industry's stud duck in Juneau." Veco diligently promoted the industry's causes through Dankworth in the legislature. Oil rewarded Veco with a steady flow of oil field contracts that in turn produced big money at campaign time.

While the promise of financial support was his main carrot, Dankworth was a master manipulator. Republican Jan Faiks, a Pennsylvania-born former schoolteacher who was elected to the senate in 1982, recalls how easy it was to fall under the Dankworth spell. Interviewed in 1991, when her lengthy political alliance with Dankworth lay in ashes, Faiks said she'd had no previous legislative experience and did not know what to expect when she went to Juneau in 1982. "I found politics is a lonely game," says Faiks. "I found quickly that no politician helps another politician. In steps Ed Dankworth. 'I want to help you,' he says. 'You are smart. I am putting the senate reorganization together. I want you to win.' "

Faiks says Dankworth called her as often as six times a day to pass on legislative gossip. He sent a car to meet her at the airport when she traveled. "He massages egos and develops an emotional dependency," she says. "Since 1983 he has put every senate organization together." When Faiks ran for reelection to office in 1986, Dankworth bestowed ultimate favors upon his protégée. Veco gave $20,100 to her winning campaign and ARCO kicked in $4,500. Her contributions from the oil and gas industry totaled $39,275, topping those of any other legislative candidate and even the Republican candidate for lieutenant governor. When Dankworth reorganized the Republican-controlled senate for the next session, he brazenly installed Faiks in the senate presidency, the most powerful office in the Alaska legislature.

Why didn't the Democrats raise a fuss over such manipulation by an outside interest? Senator Joseph P. Josephson, an Anchorage Democrat and an astute attorney who served in the state legislature from 1982 to 1988, readily admits that he went along with Dankworth's reorganizations. "Ed Dankworth functions as a broker," Josephson says. "A broker can make suggestions to senators that no senator can. His role in this situation is to make proposals. Whatever he proposes you can later turn down. He does not come selling an ideology. He is a practical person and he knows what committee appointments interest which senators, Republicans and Democrats.

"It is known that Dankworth has the ability to raise campaign money," Josephson concedes. "Perhaps a kind of mystique grows over his presence, but the appearance of power is power. I didn't want him as a friend—or an enemy." It is also known that some of the glib, statistic-laden experts the oil industry sends to Juneau to lobby on complex bills privately resent Dankworth's image as the petroleum industry's power in Juneau (and his hefty annual retainers). However, the record indicates that the leaders blessed by Dankworth served oil well.

During Jan Faiks's reign as senate president in 1987–88, newly elected governor Steve Cowper decided it was time for Alaska to close a widening oil tax loophole. He urged repeal of a concession called the Economic Limit Factor (ELF), which had been passed in the midseventies. It was a complicated formula in state law designed to give tax breaks to marginal oil fields. Part of the 1981 tax bill that eliminated separate accounting had included suspending the ELF tax break for Prudhoe. But with the senate firmly in control, the oil companies had begun reapplying it to the Prudhoe Bay and Kuparuk fields on the North Slope. Both were so large and profitable that defining them as "marginal" was laughable, but nevertheless the oil companies dealt themselves a tax break of about $150 million a year.

Cowper quickly discovered what the senate does to Alaska governors who try to tinker with rules affecting the oil industry's bottom line. The governor publicly urged a vote by the full legislature, hoping the embarrassment of protecting the oil industry in public would yield the winning legislative votes. But despite the governor's plea, ELF was blocked from even reaching the senate floor. The senate leadership bottled up the bill in committee, as they had with all recent tax bills. Jan Faiks performed right on cue. She effectively blocked the bill, much to the gratification of the legislators, who were loath to declare their stance on the matter.

Every bill of any consequence seeking to impose new environmental controls on the industry suffered a similar fate. In keeping with the high promises of safety precautions made at the time of the pipeline debate, the state at first allocated several million dollars a year for pipeline surveillance. In its 1989 survey entitled "Where Did All the Billions Go?" the Institute of Social and Economic Research at the University of Alaska Anchorage showed that the Department of Natural Resources received $4,938,000 for pipeline surveillance in 1981. As oil's political image ascended, memories of oil spill warnings faded and so did pipeline safety enforcement commitments. While budgets for public services and public payrolls swelled, funds for policing the safety of shipping oil from the pipeline shrank. They became the victim of campaign pledges to curb government spending. First, the budget covering surveillance was cut negligibly, by $300,000 in 1982. With no protest in the legislature, it was stripped to a mere $120,000 in 1983. By 1984, that category of spending was eliminated entirely.

Oil was, indeed, in charge in Alaska as the 1980s neared an end. It had the right people in the legislature to protect its revenues, and it had rendered toothless any insistence by a state surveillance agency that it live up to its safety commitments. The promises that oil would be produced in Alaska with the best environmental protection and tanker safety that money could buy were erased. Vigilance had given way to complacency, and it would take a disaster to upset the good feeling between industry and the state now. Little did Alaska realize that in yielding oil such comfort, it had been helping the industry sow the seeds for just such an event.

CHAPTER 16
Hard Aground

JUST BEFORE 5:00 A.M. on March 24, 1989, Claire Richardson, a thirty-year-old Jesuit radio volunteer temporarily working in Valdez, awoke to a hard knock on her bedroom door and station manager Dave Hammock's voice.

"There's an emergency," he shouted.

"What? No hot water?" she replied.

"This sounds serious," Hammock said. "A big tanker is aground and it's spilling oil."

Claire threw on her clothes, rushed to the phone, and called the Coast Guard. By 7:00 A.M., she was sitting before the mike of the little public radio station KCHU and delivering a news bulletin that would rumble like an aftershock across Alaska: The captain of the supertanker *Exxon Valdez* had radioed the Coast Guard sometime after midnight that his ship, carrying 1,263,000 barrels of crude oil, was hard aground on Bligh Reef, and oil from ruptured tanks was flowing into Prince William Sound.

As news of the spill was broadcast by the state's radio stations, most Alaskans reacted with disbelief. They had long been assured by oil company experts that a disaster of this magnitude would never happen. Besides, it was well known that radio stations in Alaska have virtually no news staffs and just read headlines from the papers. Yet Alaska's only morning newspaper, the *Anchorage Daily News,* had nothing about an oil spill.

Dean Fosdick, the Associated Press bureau chief in Alaska, stepped into his office in Anchorage early that morning to a ringing phone. A trusted AP member, Mark Guy of Anchorage radio station KFQD, was calling about the grounding and said the Coast Guard now was saying that 100,000 barrels of oil had already poured into the sound. Within fifteen minutes, Fosdick sent a story marked "URGENT" on the Associated Press wire. "Then," Fosdick says, "all hell broke loose." Even while Alaskans were still trying to sort out the facts from sketchy broadcasts, a media army from around the world had begun to mobilize and head toward Alaska. Nearly all made their way to Valdez. In its history the embattled little town at the head of Prince William Sound had been overrun in the gold rush, swamped by the tsunami of an earthquake measuring 8.6 on the Richter scale, and trampled by the thousands of workers streaming through in the pipeline boom. Now another event not of its own making was about to shatter its brief lapse into anonymity.

An oil disaster of world-shaking dimensions had indeed occurred in the sound. On this unusually calm March night with 10 miles of visibility, the 987-foot *Exxon Valdez*, carrying 51 million gallons of oil, struck well-marked Bligh Reef 25 miles south of Valdez, tearing huge gashes in eight of the ship's eleven cargo tanks. As the vessel foundered helplessly on the reef, 10.8 million gallons (258,000 barrels) of North Slope black crude poured into the sea.

Prince William Sound would be an environmental paradise no longer. A spreading river of oil would blacken 1,244 miles of Alaska coastline, which is about the distance of the shoreline from Cape Cod to Cape Hatteras. It would decimate marine life and close the fisheries. It would change, possibly forever, the remote, tiny villages along the inlets where Natives had lived off the sea and shore for centuries. And it would stir indignation in places far removed from Alaska. For weeks thereafter, the nation would watch television screens showing people trying to rescue scores of pitiful, oil-soaked sea otters and thousands of dying seabirds while the media kept a running death count.

Hard Aground

The oil disaster—the nation's largest ever—has been dissected by congressional inquiries, a National Transportation Safety Board (NTSB) investigation, federal and state grand juries, and a lengthy state oil spill commission hearing. Additional thousands of pages of testimony have been generated by ongoing court cases, and experts by the dozens have appeared at public forums analyzing what went wrong and speculating on the lasting impact.

Since oil began flowing down the pipeline in July 1977, tankers had made 8,858 trips to Valdez in all kinds of weather without a major accident and hauled out 6.83 billion barrels of oil without a spill at sea. What happened on the night of March 24, 1989, that caused Exxon's supertanker to land on the rocks?

The sequence of events immediately leading up to the disaster properly begins in Valdez at about noon on March 23. Joseph Hazelwood, forty-two, the moody, bearded captain of the *Exxon Valdez*, and two members of his crew went ashore for lunch and errands. Then they visited two saloons and a pizza parlor, drinking until 7:30 P.M., two hours before their ship was to leave port. No one seemed to be concerned that federal regulations bar crew members from using alcoholic beverages within four hours of sailing.

Testimony shows that Hazelwood, who resides in Huntington, New York, had a history of heavy drinking. He had entered a hospital for treatment of his alcohol problem in 1985, and had two drunken-driving convictions, the last in September 1988, only a few months before Exxon entrusted him with sailing a $120 million vessel the length of three football fields to Valdez to pick up $20 million worth of environmentally sensitive cargo. Precisely how many drinks the captain had on shore is uncertain—one crew member counted "three or four vodkas." In any event, it was enough to indicate that Exxon's troubled captain had fallen off the wagon.

A cab delivered the partying crew and a box of pizza to their ship at 8:24 P.M. The tanker left the dock an hour later, commanded by Ed Murphy, a local harbor pilot. (State law requires that licensed pilots guide the tankers through the narrow passage between Valdez Harbor and the open lanes leading to the Gulf of Alaska.) Murphy said he smelled liquor on Hazelwood's breath when they met on deck but the master appeared steady on his feet as they cruised out of the harbor.

By 11:25 P.M., Murphy had taken the *Valdez* into the open sound, turned command over to Captain Hazelwood, and left the ship in a launch. The tanker had passed well beyond the most dangerous portion of the sound, which is known as the Narrows. It could now increase speed and head due south for the open water of Prince William Sound and beyond it to the Gulf of Alaska. However, a report of floating ice from the Columbia Glacier seemed to trouble Hazelwood. The glacier in Columbia Bay is widely promoted in cruise ship brochures as a place to observe the wonders of calving ice, which is ice that breaks off from the glacier in huge pieces and falls into the sea with thunderous sounds. The ice falls into the bay about 12 miles west of the shipping lanes. Most of it melts before it can become a hazard to navigation, but sometimes the wind and current combine to send large chunks into the travel routes.

Minutes after taking command, Hazelwood called the Coast Guard monitoring station in Valdez. He said he was changing course to avoid ice reported in the normal outbound lane. Instead of slowing down the ship as many do when they encounter ice, Hazelwood chose to maintain speed and outflank the problem area. (Departing from traffic lanes is not an uncommon procedure, although the disaster of the *Torrey Canyon*—a tanker that ran aground off England in 1967 when the captain left the traffic lanes to save time—was recent enough to remind tanker skippers of the perils.) "Once we're clear, we'll give you another shout," Hazelwood said, signing off. The lone Coast Guardsman on duty in the monitoring room routinely acknowledged the change and then dropped the vessel from his radar screen. He would testify later that he felt no obligation to monitor the ship further, to check whether it was cutting corners too closely or whether its new course put it perilously close to any other navigation dangers. This attitude would later come in for sharp criticism by the NTSB.

Testimony shows that Hazelwood then put the ship on automatic pilot and went below to his cabin, leaving Third Mate Gregory Cousins, thirty-nine, in charge. His final instructions to Cousins were to start returning the vessel to the normal traffic lanes when Busby Island, which is just north of Bligh Reef, was abeam to port. However, Cousins said the captain never informed him that the vessel was on automatic pilot. By all maritime rules, Hazelwood committed a grievous dereliction of duty. Sailing in shoal waters amid floating ice floes requires the careful timing and judgment of a ship's master. This was not the time to abandon the bridge. Furthermore, he turned over command to a crew member who did not have the qualifications to conn a vessel in those waters. Not only did Cousins lack the required pilotage endorsement, but he had spent a stressful day getting the ship ready for sailing and had had only four to five hours of sleep in the previous twenty-four hours.

When the normal shift changes occurred at midnight, Seaman Robert Kagan, forty-six, took over the helm. Kagan's service record, which was produced at the NTSB hearings, showed that he had been recently graded as deficient in steering the ship even in normal seas. "He [Kagan] has been practicing steering . . . but he still requires too much supervision to be a productive crew member," Chief Mate Guy Kleess wrote in an appraisal only two months before the vessel left California for Valdez.

What occurred next can be reconstructed almost minute by minute from the NTSB testimony:

> **11:52 P.M.**—Captain Hazelwood leaves the bridge, saying he has to go below to work on messages that have to be sent before the ship leaves Prince William Sound. Cousins is now in command. The ship is traveling at about 11 or 12 knots, which is about twice as fast as its speed in passing through the Narrows.
>
> **11:53 P.M.**—Lookout Maureen L. Jones, twenty-four, a tall, rangy recent graduate of the Maine Maritime Academy just reporting on duty, sees the Busby Island light a few degrees forward on the port beam. Then she sees a red flashing buoy on the starboard bow and promptly reports the sighting to Cousins. This is Bligh Reef buoy, about 3 miles away, and it means the ship is nearing waters marked on the chart as "red sector"—dangerous water ahead.
>
> **11:55 P.M.**—Cousins, having calmly acknowledged Jones, completes a navigation fix on the chart in the wheelhouse and orders the helmsman to come right ten degrees. Cousins discovers that the automatic pilot is on. He rushes to turn it off.

11:57 P.M.—Ship is not responding. Lookout Maureen Jones again enters the wheelhouse and reports that the Bligh Reef light is still flashing on the starboard bow. For the ship to be in safe water, the warning buoy should be on its left, which is the port bow. The vessel is now in dangerous water. Cousins increases rudder order to right twenty degrees.

11:59 P.M.—"I observed the Bligh Reef buoy light," Cousins later recalled. "I did not like what I was seeing. There was very little transfer off our original track. I ordered 'Hard right!' " Helmsman Kagan's version: "Cousins grabbed the wheel. We both spun it together."

12:05 A.M.—Cousins phones the captain in his cabin below. "I believe we are in serious trouble," he tells him.

12:09 A.M.—The tanker's hull starts scraping the jagged rocks on Bligh Reef. Cousins later recalled that he felt an initial shock at the end of the conversation. "First a slight roll," he said, "then a series of jolts—six or seven very hard jolts." Now the ship is fast aground. Cousins waited too long before he ordered the first turn to the right. The subsequent sharp turn to the right had come too late.

Chief Mate James R. Kunkel, who was asleep after a long day of supervising the loading operation, was jolted from his bunk as the vessel shuddered and clanged. He dressed quickly and dashed up to the bridge. A distraught Cousins told him the ship was aground and the captain was on the phone talking to the Coast Guard.

Kunkel rushed down to the cargo control room to see how fast the ship was losing oil. The gauges were a blur. The center and starboard cargo tanks were discharging so rapidly that at least 115,000 barrels already had been lost. The stability of the vessel was now a serious concern. "I don't mind telling you that at that moment I knew my world would never be the same again. I feared for my life," Kunkel told the NTSB hearing board. Using the vessel's load master computer, Kunkel got a fast printout of the stresses. It told him the ship was in considerable danger of capsizing. Running to deliver the reading to the bridge, he noticed strong cargo vapors in the passageways. His concerns for the safety of the crew multiplied.

"Shall I ring the general alarm?" he asked as he approached Hazelwood, who seemed deep in thought on the bridge.

"No, Jim, stay cool," Hazelwood replied. "Let's get a good assessment before we fly off the handle."

Kunkel testified that he informed the captain that computer calculations showed the tanker "was not stable to move." Nevertheless, Hazelwood restarted the engines and tried to dislodge the ship off the reef.

Engine room logs indicate the captain gave six successive speed commands before deciding the vessel couldn't be moved. (Had he succeeded in getting the ship off the shoals, it would have capsized and sunk, making the record-breaking disaster many times worse, maritime experts say.)

At 2:30 A.M., two top officers of the Coast Guard's Valdez Marine Safety Office were roused from their beds and boarded the pilot vessel *Silver Bullet* along with Dan Lawn, the state's environmental supervisor in Valdez. They sped toward the stricken tanker. Ironically, the closest thing to human tragedy in the entire hectic night occurred on the way to the ship. A female deckhand on the *Silver Bullet* was told to keep an eye out the port window for ice. As she opened the window, her hair caught in the electric window fan. She avoided being scalped only by reaching for a knife and cutting off the trapped strands of her hair at the roots.

The boat approached the tanker, on the starboard side, which was the leeward or protected side of the ship. But when the boat's light shone upon the waters, the crew saw the sea boiling with oil, as high as 16 inches above the surface. The crew had difficulty moving the ladder to a safe spot where they could make the long climb up the side of the tanker.

Chief Warrant Officer Mark J. Delozier, a member of the Coast Guard boarding party, testified that he found the fumes so thick on coming aboard that he feared a fire or explosion. Yet, he added, he found Captain Hazelwood on deck drinking coffee and smoking. "That's not a prudent thing to do, Captain," Delozier said. Hazelwood put out his cigarette. When he got within four feet, Delozier detected alcohol on the captain's breath. "He smelled like someone who had been sitting in a bar and drinking for some time," Delozier testified.

Delozier asked Lieutenant Commander Thomas Falkenstein, his executive officer, who had boarded with him, to step aside. After conferring briefly, the two Coast Guardsmen decided that the captain and crew should be tested for alcohol, but they were not sure who should do it. (This procedural uncertainty would also be sharply criticized later by the safety board.) Finally they decided to call the Alaska state troopers. A state policeman arrived shortly after 6:00 A.M., but he carried no equipment to conduct any testing or to collect urine samples for subsequent testing. "Geez, I thought you had some kind of wild man on board. I didn't come to give an alcohol test," the trooper was quoted as saying.

Next, the Coast Guardsmen called their base at Valdez for a medical technician to collect toxicological samples. By coincidence, a technician from Anchorage had been visiting Valdez. He was tracked down at the airport and told to report to the ship with equipment to take blood and urine samples. When he arrived, the technician discovered that kits for obtaining toxicological samples had been on board all the time as a standard part of the tanker's equipment. By now, ten hours had expired since the grounding.

Hazelwood's blood alcohol content tested at 0.061 percent, well above the 0.04 percent allowed for vessel masters by Coast Guard regulations. The NTSB findings projected that, given the elapse of time, the captain had an estimated blood alcohol content of 0.2 percent at the time of the grounding (most states prosecute for drunken driving at 0.1). The findings suggested that the captain drank on board after the ship was under way or after the grounding.

The NTSB listed forty-seven findings in its report on the accident. It concluded that the *Exxon Valdez* hit the rocks at Bligh Reef because of errors on the bridge. Cousins, it said, did not act quickly enough to change course and permitted the ship to travel a full mile into dangerous water toward the reef before giving the first order to turn right. The hearings attributed the errors to an alcohol-impaired captain, an unqualified and overworked third mate, and inadequate training of personnel. Such negligent management might have been expected in an accident involving a bootstrap company running a banana boat. But this was the crew assigned to a state-of-the-art tanker owned by one of the richest corporations in the world.

How did the corporation's tanker operations sink so low? The

NTSB placed heavy blame on Exxon Shipping Company, not only for inadequately monitoring alcohol abuse by employees but also for "pursuing reduced manning procedures." The policies "do not adequately consider the increase in workload caused by reduced manning," it found. The company "had incentives and work requirements that could be conducive to fatigue." And it "had manipulated shipboard reporting of crew overtime information that was to be submitted to the Coast Guard for its assessments of workloads on some tankships."

Though it was small satisfaction for ex–safety crusaders like John Rader and Chancy Croft, the NTSB report also found that "if the Exxon Valdez had been fitted with an 11-foot double bottom, the resulting oil loss would have been small, and possibly eliminated."

Exxon's corporate leaders accepted the safety board's criticism with pledges of reform, but no axes fell at the corporate level. Captain Joseph Hazelwood became the fall guy. Exxon fired him as soon as the results of the alcohol tests were announced, and the Coast Guard lifted his license for nine months. Alaska authorities filed three criminal charges against him and a state grand jury indicted him on a felony. For a while, anger at Hazelwood rose close to a lynching pitch. He traveled under heavy security. On a return flight to New York, fellow passengers mocked him. But soon, he simply became the butt of jokes. By the time he returned to Anchorage for trial, in February 1990, he was viewed as almost a tragic figure.

Hazelwood's lawyers argued that under federal law the captain was immune from prosecution because he had called the Coast Guard and reported the spill as soon as it happened. They contended government policy considers it more important to have spills reported promptly than to prosecute the spiller, and sought to have the case dismissed. But superior court judge Karl Johnstone ordered the trial to go on, on the grounds that the huge spill would inevitably have been discovered whether or not the captain reported it. Dressed in a conservative blue suit, Hazelwood sat between his attorneys, his deep, dark eyes staring forlornly into space and his fingers twisting the end of his short, black beard as prosecutors called forty-six witnesses. Several testified about his drinking and on-board parties.

But in the end the jury of nine women and six men were persuaded Hazelwood was a scapegoat for Exxon's crimes against Alaska. They acquitted him on all serious charges involving drinking and felonious damage of property. They found him guilty only of one misdemeanor—negligent discharge of oil. Judge Johnstone attempted to temper the unexpectedly mild verdict by ordering Hazelwood to pay $50,000 in restitution to Alaska and sentencing him to one thousand hours of cleanup duty. The captain appealed, and on July 10, 1992, the conviction was overturned by the Alaska Court of Appeals on the very grounds that Judge Johnstone had denied: The federal government does offer immunity from prosecution for anyone reporting an oil spill.

Hazelwood received the news in New York, where he teaches at New York Maritime College, training student sailors at sea. The possibility that he will ever return to Alaska to clean a single oil-stained rock now seems remote. Alaska is appealing the overturning of his conviction to the state supreme court, but by the time the legal skirmishing runs its course, years of tide and rain will have made his sentence moot.

CHAPTER 17

Mopping Up

GOVERNOR STEVE COWPER was known in Alaska as "the high plains drifter," and not only for the cowboy hats and western boots he favored. A Democrat elected in 1986, he quickly became a loner in the state capital. He refused to kowtow to the legislature, and he wanted oil to pay a fairer share of taxes to the state. As a result, he was steadily trampled by both. Though respected for his integrity, the governor had nothing but bad luck from the day he took office. The state legislature cut the heart out of his pet project for Alaska—a permanent endowment for public education. Then the oil lobbyists persuaded the state senate to bottle up his Economic Limit Factor (ELF) bill, which would have closed a tax loophole exploited by oil firms.

As though unable to shake a losing streak, Cowper picked the fateful day of March 24, 1989, to announce that he wasn't going to take it anymore. In an unusual 7:00 A.M. press conference in Fairbanks, Cooper declared that even though he still had twenty-one more months to serve, he would not seek reelection. He was braced for a flurry of questions, but not the one posed that morning by Sam Bishop, political reporter for the *Fairbanks News-Miner*.

"Governor, what are you doing about the oil spill?" Bishop asked.

Oil spill? The governor's face reddened. He had spent the night virtually incommunicado in his Fairbanks home, which has an unlisted phone number. He had no inkling until now that the worst oil disaster in North American history was to join the list of indignities heaped on his administration.

Cowper had planned to hold a teleconference at noon with his cabinet and other close aides to explain why he would not run again. Instead, he left Fairbanks abruptly and flew to Prince William Sound. A small boat met his float plane, and by midafternoon Cowper and Dennis Kelso, commissioner of the state's Department of Environmental Conservation (DEC), climbed a long, swinging chain ladder onto the crippled tanker.

"I tried to pay my respects to the captain," Cowper says. " 'The captain is not here,' I was told. If the captain was down below, it was clear he wouldn't come up to talk to me." The governor spotted Dan Lawn, DEC manager of the Valdez office, checking the oil loss with the chief mate. He grabbed Lawn by the arm and led him off to the wing bridge.

"Tell me, what's going on?" Cowper asked.

"Nobody is cleaning anything up," Lawn replied. He told the governor that at daylight he had called the Alyeska Pipeline Service Company, the consortium set up by the seven big oil companies in Alaska to transport oil, and asked where the cleanup barge was. It was urgent, he said, that they immediately muster every piece of cleaning equipment they could find. "They told me the boat was on the way," Lawn said.

The oil spill contingency plan worked out with the state and Coast Guard called for Alyeska to have a barge with booms and skimmers on the site in five hours. Nearly ten hours had passed by now, and as Lawn later discovered, the equipment was still sitting in Valdez. Raymond Cesarini, manager of the Sea Hawk Seafoods processing company, who flew over the entire spill area at two o'clock that afternoon—almost fourteen hours after the accident—told the press, "I couldn't see even a Kleenex in the water to clean up oil out there."

More than 10 million gallons of oil had already poured into the sound by midafternoon. About 42.2 million gallons were still on board, and the ship was in danger of capsizing. Exxon managed to divert an inbound tanker, the *Exxon Baton Rouge*, to take on the oil from the stricken ship, but tense hours were spent in search of the fenders needed for the tanker to come alongside. They were finally located in Valdez under 4 feet of snow. Fortunately, calm weather prevailed until a crew could dig them out and rush them to the tanker 25 miles away.

The good weather continued for three days—most unusual for March in Prince William Sound. The calm was ideal for booming and skimming and at least blunting the leading edge of the river of oil. But from the outset the cleanup stuttered in a not-so-funny comedy of errors, the first crucial days wasted in bureaucratic posturing, bumbling, and feuding between industry and the regulatory agencies.

Alyeska was unable to dispatch the containment booms and skimmers in time to attack the leading edge of the oil because the emergency response barge had not been loaded. When a crew finally dug out the equipment, only one worker of the fifty assembled for the response team knew how to operate both the forklift and the crane needed to load the devices. He had to run from one machine to the other. It took fourteen hours before the barge went into action. By nightfall, only 7,500 feet of boom had been laid. Observers flying above described the booms as matchsticks trying to hold back a black bore tide. The spill response plan had called for Alyeska to have 20,000 feet of boom at the ready.

Exxon, which stepped in to direct the cleanup, concluded that the spreading oil could not be contained by mechanical means and that chemical dispersants were necessary. When it asked the Coast Guard for permission to use Exxon's proprietary compound COREXIT 9527 to pulverize the oil, the commander said he had to consult with state and federal environmental officials who had previously expressed concern about the chemical's possible impact on fish. When permission finally came, test treatments were so crudely applied that they were predictably ineffective. (The dispersants were dumped from buckets mounted underneath a helicopter and much of the spray rained down on Coast Guardsmen watching from the deck of the tanker.) It turned out that the Coast Guard in fact had the authority to permit dispersants at the outset.

Exxon next considered burning the oil. The state balked, fearing legal repercussions. It finally issued a permit "only if burning did not impact public health or violate air quality standards." Experts say burning is the best way to deal with an oil spill early in the cleanup process. The unrealistic conditions of the permit kept Exxon from proceeding quickly, and when a test burn was at last tried, it, too, was badly managed. No one had notified the Indian village of Tatitlek downwind, and its alarmed inhabitants were caught in clouds of choking smoke. By the time cleanup leaders regrouped to pick the site of the next burn, the oil had been so diluted with water that it would not ignite.

On the third day following the spill, Exxon brought in a specialist to meet with fishermen, villagers, and environmentalists to identify critical habitat areas that needed priority protection. Exxon wanted the meeting to be held in a bedroom of the Westmark Hotel in Valdez and closed to the public. But it had not counted on Riki Ott, a Cordova fishing boat owner with a doctorate in marine toxicology who had predicted a major spill in a hearing on the very eve of the accident and now used her newly won renown to call for candor on the part of the oil company. Exxon grudgingly moved the meeting and opened it to the public. "Many salmon creeks and the herring hatchery area were identified, but the next day the wind blew and trashed everything," Ott says. "We then just pooled our forces to try to save the hatcheries."

Gale winds got the oil slick moving at 30 miles per hour, "as though it was on a superhighway," Frank Iarossi, president of Exxon Shipping, said at the day's briefing. Within a week, the oil track had moved 90 miles down the sound, blackening shorelines and suffocating estuaries along the way. Fewer than 7,000 of the 258,000 barrels had been mopped up. The spill was now out of control.

Iarossi had arrived at the scene by the afternoon of the catastrophe, accompanied by three public relations men and an attorney. Their mission was damage control. Besides mopping up the spill, Exxon knew it had a massive job on its hands to refurbish its stained corporate image. "We will pick up, one way or another, all the oil that is out there," Don Cornett, Exxon's public affairs manager, promised in a press release immediately after the spill. "We hope to leave Prince William Sound the way we found it." This was incredible public relations puffery, as was obvious to any reasonable person looking at the blackened bay. Exxon whisked Cornett out of Alaska, but his hyperbole haunts the corporation in every assessment of the cleanup to this day.

Exxon set out to restore Prince William Sound and its own image by attacking both problems with the weapon it knew how to use best—money. It brought in 160 spill experts and managers, flew in thirty-seven cargo planes of equipment, and gave a blank check to Veco International, the service company run by Bill Allen, the industry's champion in Alaska. Allen promptly spread the money around. Veco hired nearly half of the local fishing fleet to collect oil. The firm put thousands of workers—with preference to Alaskans—on the payroll to clean up the beaches. While much of the nation languished in a recession, Exxon created an economic boom in Alaska. The news that people were getting $16.69 an hour just to scrub rocks brought job seekers streaming into Valdez. All-but-dormant union halls suddenly bustled with hiring activity. Fortune seekers came from everywhere to cash in on the boom any way they could—and crime rates soared accordingly.

Cordova fishermen, who dominate the fishing industry in Prince William Sound, had been the earliest and loudest critics of the oil companies' intrusion. They had opposed the pipeline and warned of oil disasters from the beginning. When news of the spill hit the harbor, angry men and women poured onto the docks. Some raced into the bay with buckets, nets, even mops. But the outrage galvanized into concerted action when it dawned on the community that the salmon hatchery at Port San Juan, the lifeblood of the Cordova fishery, could be wiped out by the spill. The fishermen appealed to Alyeska for help. They were told, in effect, to file a claim. "Alyeska said there is no way we can save the hatchery," recalls Tom Copeland, a Cordova boat captain who has fished the sound for twenty-eight years. "We said we will save the hatchery."

The Cordova District Fishermen United quickly voted to spend a million dollars, if necessary, and the city pledged a half million to fund the frantic rescue operation. "We called all over the world for the booms," Copeland says. "We had the booms in place in six days after the state told us none were available. And we had a hundred fishing boats ready in a week to lay the booms and keep them in place when the bay kicked up."

Copeland's partner, Floyd Hutchins, found out that VHF radio in Valdez, which Exxon used to coordinate cleanup operations, was unable to reach boats heading for the hatchery. Hutchins anchored his boat in the bay and became a relay station between Valdez and the hatchery. The confused voices out of Valdez further frightened the Cordova fishermen. "After listening for three days," Copeland says, "Floyd knew that no one up there knew what he was doing." Exxon was paying for expensive skimmers to pick up oil, but they were losing the battle. Of the forty-seven skimmers listed as deployed one day, only about half were working at the same time. Some broke down, some were immobilized by rough water, and others were just waiting hours in line to dump their cargo into a barge.

Copeland felt that something more had to be done. On Day Eighteen after the spill, he created his own skimming team. He drove his seine boat into the bay to intercept the oil coming down the sound. With the help of two other Cordova fisherman manning rubber rafts, they began scooping oil into five-gallon herring roe cans. "There were plenty of empty herring roe cans around Cordova—about thirty thousand of them. The state canceled the herring season because of the spill and no one was filling those cans with anything else. . . . I told Floyd to radio the Exxon bosses in Valdez that I will show up with a hundred buckets of oil, and they better have a plan [to accept it]."

Her decks loaded with filled herring cans, the 47-foot *Janice N.* sailed into Valdez and tied up at the town dock. The boat and the men on board were covered with oil. A pack of reporters covering the spill rushed to the pier, their cameras clicking. Copeland made the most of the moment. "The Japanese would be paying us a hundred forty dollars a bucket for roe kelp which we couldn't catch because of the spill," he told Exxon officials on the scene. "Give us a twenty-dollar-a-gallon bounty on oil and we'll clean it all up."

Chagrined Exxon officials were forced to make a deal. They paid $5 a gallon plus a half-rate contract of $1,500 to cover fuel and groceries for one boat. "At first they put a limit of two thousand gallons a day," Copeland said. "They must have been terrified that I had a whole bayful of oil hidden somewhere. But the next day they withdrew the limits and even issued five more boats half-rate contracts." With the spill in its third week, newspapers reported that three Cordova fishermen picking up oil in buckets were doing about as well at cleaning up Prince William Sound as the elaborate skimmers used by Exxon. Copeland's team would boom up a pool of oil and a man in a Zodiac inflatable boat would be placed in the middle on his knees to scoop the oil and pipe it aboard the *Janice N.* The boat collected 1,500 gallons a day, a remarkable feat since Exxon's entire skimmer fleet averaged about 2,180 gallons day.

"Around the end of the third trip," Copeland said, "Red Starr, an independent contractor, showed up with a sewage vacuum truck which had a super sucker. He hung around for six days and couldn't get a contract. I finally went up to see Ed Bechtel, a contract coordinator for Exxon, and told him I will go to the press and create a nuisance. I'd tell them about the ninety-foot crab boat earning ten thousand dollars a day and still waiting for something to do while you refuse to hire the guy with the best equipment."

"Promise good PR?" Copeland was asked.

"I promise, I promise," he replied.

"We had four good days with the super sucker," Copeland said. "We picked up forty-four thousand gallons on five boats. Had we gotten out sooner, I believe the fishing fleet could have caught the entire spill. We had the pumps, boats, tankage. Those vacuum fish pumps in our world-class herring fleet could have pumped oil the way we pump herring out of the nets. This fleet scoops up eight thousand tons of herring in a twenty-minute opener. We never had the cooperation of government and we were told by Exxon that we couldn't do it because we never tried it on oil."

The failure to make greater use of the Cordova fishermen to save their own bay is still a controversial topic. However, their success in saving the Port San Juan hatchery—after being told it couldn't be done—was an epic triumph in the oil spill. Exxon later presented a check for $300,000 to the enterprising fishermen and managed to claim a share of the publicity. In the meantime, the company proceeded to pacify the fishermen in substantial ways. Veco paid the owners of hundreds of fishing boats between $3,000 and $10,000 a day each to ferry workers, string booms, and pick up garbage. Some boats, such as the costly crabber referred to by Copeland, sat in port for days at a time drawing full contract pay while waiting for instructions. Crews on boats under hire were not allowed to talk to the press or permit reporters aboard, a condition the boat owners happily accepted, since many were pocketing far more money than they could have made fishing. In Valdez Harbor, they became known as "spillionaires."

As part of the image-repair process, Exxon also spent lavishly to appease environmentalists and animal lovers. Oiled sea otters, which seemed to be part of every televised news report on the spill, were plucked from sticky beaches and sent to special treatment centers, where they dined on crab and shellfish specially flown in during their much-publicized rehabilitation. An estimated 250 otters were saved—at a cost of about $90,000 apiece. Many Alaska fishermen grumbled at the expense. While lovable in the eyes of tourists, the protected sea otters rob crab traps and deplete shellfish beds, and it is common to hear fishermen boast of shooting them when no one is watching.

The total cleanup force grew to about fifteen thousand workers, not including the hundreds of volunteers who cleaned the blackened beaches. While most of the paid workers were Alaskans grateful for the jobs, the easy money also attracted frauds and loafers. Connie Goss, a Valdez housewife, scrubbed rocks with absorbent cloths the entire summer. "I had a hard time getting a job because so many poured into town from the Outside. They would pick an address and phone number out of the phone book and list that on their application to show they were local. Then they would go down to the hiring hall each morning and wait for their name to be called." The disparity in commitment created disharmony on the beaches. "Goofing off and bitching," Mrs. Goss adds. "I never heard so much ungrateful bitching in my life, and it miffed us that the goof-offs hardly ever got fired. Some left after a few weeks saying they couldn't face scrubbing another rock. Some wise guys who lied to get jobs took one look at the oily mess and lasted only one day."

Nevertheless, Exxon and Veco won general acclaim for mounting an incredibly complicated logistical and engineering operation that spared no expense. Their flotilla of omnisweepers and maxi-barges was awesome. Working in tandem with the rock scrubbers, these state-of-the-art machines blasted oil out of the beaches with high-pressure hoses that shot scorching water. Perils abounded, even in such daily routines as landing and removing workers from the beaches in 40-knot winds and battering seas. Some boats were smashed, but not a life was lost.

Almost immediately, Exxon paid more than $300 million in undisputed claims. Some went to fishermen who lost out on the spill-curtailed fishing season and did not get cleanup jobs. Substantial sums were paid to Native villagers whose subsistence way of life was disrupted when, in the words of one elder, "white man made the water die." These payments are expected to pale in comparison to what Exxon will have to pay to settle the disputed claims in the hundreds of lawsuits still pending.

Exxon intended to end the cleanup after one summer. But it seemed that whenever it declared a beach cleaned, an official from the state's DEC would lead a press tour to the beach and find soiled rocks or clumps of oil. The official would point out that this was not what Exxon had promised in its pledge "to pick up all the oil that was out there." A tug-of-war for media approval grew in intensity as the summer wore on. Exxon rented dozens of rooms at the Westmark Hotel in Valdez, where it held regular press conferences and gave regular updates on beaches cleaned and animals saved. The DEC, meanwhile, leased every room but the bar at the American Legion hall a few blocks away. It gave a running account of the death toll, which by the end of the first summer amounted to 980 sea otters, 138 bald eagles, and 33,125 seabirds. The actual figures were probably much higher, the state would point out, because the tally counted only the actual carcasses placed in the freezer.

Exxon returned to clean for a second summer in 1990. This time it came equipped with a new semantic arsenal in the battle for the public mind. No longer did Exxon refer to beaches as "cleaned"; the term now was "treated." Crews on the beaches were no longer scrubbing rocks, but engaged in "treatment for purposes of environmental stabilization." The state countered by referring to the rock-scrubbing process as Exxon's attempt to "remove gross contamination." And the U.S. Fish and Wildlife Service was not to be outdone. Killing seals in order to inspect them for contamination was called "a collection." The boats that collected dead animals were termed "rescue boats."

By the end of the second summer, thousands of tons of oily waste had been scraped off the beaches. Much of it was packed out to incinerators outside of Valdez. Hundreds of tons of waste that couldn't be burned were loaded on barges and shipped to a hazardous-waste dump in Arlington, Oregon. But debates were developing about whether the intensive cleanup was doing Prince William Sound more harm than good. The once-tranquil shoreline and islands had been pounded by constant traffic and industrial activity. An impact study by the National Oceanic and Atmospheric Administration found that scouring beaches with hot water applied with high pressure essentially cooked sensitive marine life. The agency concluded that, biologically speaking, beaches left untreated were in better shape. Meanwhile, Natives began complaining that spill workers were desecrating once-secret ancient burial grounds and stealing artifacts from anthropological sites.

Alaskans today are thoroughly confused over the effectiveness of the cleanup. About the only thing they know for certain is that Exxon spent a lot of money on the operation—the common understanding has grown to $2.5 billion, although even Exxon critics concede it could be more. But if they have been listening to the findings of the various investigations into the spill, they may also be aware that for years the oil industry cut costs at the expense of protecting Alaska—and that the state's top environmental enforcement officers were aware of the backsliding but lacked the political courage to hold the oil industry accountable for its commitments.

Several federal hearings on the spill were held, but the most intensive was a state inquiry conducted by the Alaska Oil Spill Commission, an independent panel appointed by the legislature. The spill would not have occurred, the commission concluded on January 5, 1990, "if Alaskans, state and federal governments, the oil industry, and the American public had insisted on stringent safeguards. It could have been prevented if the vigilance that accompanied construction of the pipeline in the 1970s had been continued in the 1980s. The original rules were consistently violated primarily to ensure that tankers passing through Prince William Sound did not lose time by slowing down for ice or waiting for winds to abate."

The commission's chairman, Walter B. Parker, had headed the task force that had laid the ground rules for marine operations back in 1977, when tankers first began loading at Valdez. Couched in his terse covering letter was a sense of betrayal by the greedy oil industry and the unwilling regulators. "Concern for profits in the 1980s obliterated the concern for safe operations that existed in 1977," Parker wrote.

While the commission submitted a comprehensive policy to minimize chances of future spills, it did not go into the details of how the risk-taking oil industry managed to lull state and federal enforcement agencies into docility. That story emerges in conversation with Dan Lawn, the crusty, bearded Alaskan in the Valdez office of the DEC.

Fresh out of the University of California with a degree in environmental resources engineering, Lawn came to Alaska in 1973 at the age of twenty-seven. Lawn was part of a new generation of technicians who were fusing their technological training with political activism. Often with the pro bono help of environmentally sensitive lawyers, they prodded business and government on their obligations to the earth. Their growing power frequently put government regulators in conflict with environmentalists on one hand and private enterprise on the other.

Lawn worked on the start-up team that built the oil consortium terminal at Valdez. Shortly after oil began flowing in 1977, he joined Alaska's new Department of Environmental Conservation as an environmental engineer to help oversee the terminal operations. DEC was to set discharge limits, enforce oil spill prevention, and investigate pollution complaints.

Lawn had been well aware that L. R. Beynon, British Petroleum's top pollution specialist, had testified before a congressional committee in 1971 that Alyeska would handle spills of any size, and that "the best equipment, materials, and expertise, which will be made part of the oil spill contingency plan, will make operations in Prince William Sound the safest in the world." For the first few years, Alyeska, which is 50 percent owned by British Petroleum, seemed intent on complying with those high-minded promises. It put in place a spill response team manned around the clock by a force of twenty-four full-time people, much like a fire station. Tanker lanes into Valdez were set to ensure safe travel, restrictions were imposed to limit operations in high winds, and agreements were even reached on slowing down speeds of tankers that encountered ice.

As the largest partner in Alyeska, BP had the clout to see that its promises were kept. And the company supports a model of oil spill preparedness at the Sullom Voe oil terminal in Scotland's Shetland Islands. The shipping center for North Sea oil has similar navigational hazards, but is subject to at least thirty more restrictive regulations governing spill response and tanker safety, as Jonathan Wills points out in his book *A Place in the Sun*. It may be that the Shetland Islands has a government courageous enough to keep the oil industry's feet to the fire. Or perhaps in remote, undeveloped Alaska, BP felt it could get away with lowering its standards. In any event, in 1979 the oil industry won the court case overturning the oil tanker oversight and regulatory system initiated by the forceful legislative coalition led by Chancy Croft. When the state did not appeal, shippers stopped following the rules and the Coast Guard decided to stop enforcing them, the oil spill commission concluded.

It soon became apparent to Lawn in Valdez that the oil consortium had begun to renege on its spill prevention promises. He saw Alyeska's spill response team shrink from twenty-four to twelve. People trained in fighting oil spills were assigned to other duties, mainly the labor force. When Lawn asked why Alyeska now sent so few people to clean up recurring tanker loading spills, he said he was told only, "Lawn, you're just negative."

Alarmed, Lawn wrote a lengthy memorandum to his superiors in May 1984, stating that he was witnessing "a general disembowelling of the Alyeska Valdez Marine Terminal operation plan." Besides the reductions in the spill response team, he listed seventeen other problem areas at the terminal, nearly all involving broken promises by the oil industry. Seven months later, he followed up with a ten-page letter identifying more commitments being broken at Alyeska, including failures to make improvements in the deteriorating ballast water treatment system, which was suspected of dumping oily wastes into the bay. He complained about the cuts in marine personnel even as tanker shipping increased. He warned that the reduction of oil spill cleanup booms had reached "unacceptable levels."

By the end of 1984, the terminal was loading fifty tankers a month, compared to thirty-four a month when it had opened. Small spills were becoming more frequent, cleanups were increasingly ineffectual, and cleanup drills were almost a farce. In a drill conducted in front of federal observers in 1984, the main boom drifted under a tanker and sank, and a workboat didn't have the power to pull it out of the water.

When the state began to investigate the Prince William Sound allegations, Alyeska suddenly tightened admittance to the terminal. "The more we found, the more difficult access to the terminal became," Lawn testified. "Restrictions delayed entry to a point that on some occasions field staff were unable to carry out inspections." Alyeska countered criticisms with its own studies and managed to sidetrack a budding investigation by the state attorney general's office, where many of Lawn's complaints landed and died.

Lawn was advised by superiors that a better way to obtain compliance was to tighten regulations in the spill contingency plan that Alyeska must renew with the state every five years. But in 1987, while DEC technicians protested that the state was "giving away the store," incoming commissioner Dennis Kelso signed a still toothless contingency plan without reading it and without bothering to visit Valdez and hear out the angry staff. The state did tell Alyeska that its approval could be withdrawn if Alyeska didn't do better on spill response, but the warning rang hollow. An Alyeska spokesman publicly branded Lawn "a jerk and a troublemaker." The company attacked the credentials of those inside and outside of the department who expressed agreement with his criticisms. At least three highly regarded professionals in the department refused to be hounded, and left.

Lawn continued to write reports about Alyeska's pollution violations and its inadequate spill preparedness, even though he was convinced no one read them, or cared. "DEC's budget kept going up, but funds for surveillance kept going down," Lawn says. "We were reduced to a paper tiger." But he refused to quit, largely because others—some even within Alyeska—joined his battle. In fact, a virtual underground of scientists, environmental activists, and concerned citizens developed an information pipeline that delivered accounts of Alyeska pollution violations to the office of U.S. Representative George Miller (D-Calif.), a frequent critic of Alyeska and of the oil industry's expansion into Alaskan wilderness.

The unlikely principal conduit in this operation was Charles Hamel of distant Alexandria, Virginia. Hamel says the oil companies drove him out of the shipping business when he questioned why there was so much water in the crude oil he brokered. Exposing industry hypocrisy on environmental matters was his way of getting even, he concedes. Hamel's activity surfaced in 1991 when Alyeska admitted it had spent $287,000 to hire Wackenhut, a Florida-based security firm, to spy on Hamel, presumably to find out who was leaking information. (Wackenhut operatives set up a phony environmental firm to gain Hamel's confidence and searched his garbage, phone bills, and living room for clues.)

Alyeska's hapless performance in the *Exxon Valdez* oil spill should have vindicated Dan Lawn. But not in the eyes of his bosses at DEC. In fact, the state took Lawn off Alyeska surveillance, moved his desk into the back of a DEC storeroom at Valdez, and assigned him a vague title with vague duties. In the public uproar that followed, DEC officials tried to deny that Lawn had been demoted.

Whether Dan Lawn and the other whistle-blowers will be fully vindicated remains to be seen. A complete accounting of the cost of the oil spill in Prince William Sound is years away. The oil industry has already paid dearly in cash, and hundreds of lawsuits are still pending. The long-range damage to the oil industry's credibility may also be costly. The spill occurred just as the oil companies and the state of Alaska seemed to be gaining the upper hand in persuading Congress to open the protected Arctic National Wildlife Refuge for drilling. They now must overcome a new arsenal of arguments, the most pointed being "Can you believe anything the oil industry promises?"

Unfortunately, there is no question that the damage to Prince William Sound is lasting. Sixty-seven percent of the 10.8 million spilled gallons was never recovered. Was the spill the national disaster at first feared? The full answer will take years to emerge. But the sound will never be the same again.

CHAPTER 18
Day of Reckoning

A PHALANX of smartly attired lawyers, followed by a dowdy horde of reporters, marched into federal courtroom number 2 in Anchorage on April 25, 1991. They had come for Exxon's day of reckoning on the crimes it had committed against Alaska in spilling 10.8 million gallons of oil into Prince William Sound. On February 27, 1990, a federal grand jury in Anchorage had indicted Exxon Corporation and its subidiary, Exxon Shipping, returning five criminal charges against each. Two felony counts charged that both had violated the Dangerous Cargo Act by employing crew members who were "physically and mentally incapable of performing duties assigned to them," and that each failed to make sure that the wheelhouse of the *Exxon Valdez* was "constantly manned" by competent crew. The three misdemeanors accused them of polluting in violation of the Clean Water Act and the Refuse Act and killing birds in violation of the Migratory Bird Treaty Act. Though no individuals were indicted, U.S. Attorney General Dick Thornburgh said in Washington that the government was "throwing the environmental book at Exxon" and estimated that maximum fines could reach $600 million. Back in

Alaska, Governor Steve Cowper found the criminal charges interesting, but said his real concern was how much Exxon was going to pay the state to repair the damage to Prince William Sound.

Now Judge H. Russell Holland had before him for approval a $1 billion criminal and civil plea bargain agreement. Of the bundle, $100 million was a criminal fine and $900 million was to be used for restoration of the sound. It was the largest recovery ever proposed in an environmental enforcement case. Yet the magnitude of the catastrophe itself had no precedent. And the outpouring of public comment told the judge that Alaskans were troubled about settling with Exxon without a trial.

Ever since the supertanker *Exxon Valdez* struck Bligh Reef two years before, legal scholars had debated—and Alaskans had wondered—how to place a dollar value on the deaths of countless seabirds, mammals, and marine organisms; the destruction of the sound's delicate intertidal zones; the saturation of hundreds of miles of rocky beaches, blasted clean on the surface but hiding untold depths of oil; or the unknown future effect on commercial fishing, the sound's lifeblood. Some argued that it is impossible to quantify such losses in dollars and cents, but society offers no other way. And the U.S. Justice Department wanted to strike a deal with Exxon as early as possible.

In 1990, the company had nearly succeeded in settling on the cheap. Maintaining that Prince William Sound was recovering rapidly after a year's cleanup, the company agreed to admit minor criminal violations and pay $550 million in civil damages to cover restoration. The U.S. Justice Department, at the urging of Attorney General Dick Thornburgh, seemed eager to accept the deal rather than risk defeat in court.

But Alaska balked. Governor Cowper, then in his final months in office, declared that $550 million was not enough to restore the sound. Besides, the settlement had a sweetheart clause according to which the Justice Department agreed to delay civil suits against Exxon for four years. That would have robbed Alaska of essential federal help in such suits. Stung by the state's rebuke, the federal negotiators simply threw up their hands. "Thornburgh wouldn't work with us after that," Cowper says.

But in 1990, seventy-one-year-old Walter Hickel rose from political ashes to again become governor of the state, bringing with him a long history of rapport with the oil industry. In his first month in office, Hickel invited his friend Lawrence Rawl, chairman of Exxon, to Juneau. Two months after the meeting, Hickel produced the $1 billion settlement that was now before Judge Holland. The judge, a Republican appointed to the bench by President Reagan, had been fully expected to rubber-stamp the deal. But three days before the deadline for public comment, the first report of a federal-state scientific damage assessment study had been unexpectedly filed in federal court. The eighteen-page report summarized the preliminary findings of about $70 million worth of scientific work commissioned soon after the spill. It contradicted Exxon's persistent glowing portrayals of a fast recovery, revealing that the damage to birds and mammals was far more lethal, enduring, and pervasive than the public had been led to believe. The Hickel legal team had strenuously objected to the release of this information, claiming it would expose their hand should the Exxon case go to trial. But Paul Gertler, a deputy director of the U.S. Fish and Wildlife Service for the Alaska region who chaired the damage assessment management team, successfully fought for disclosure on the grounds that the public was entitled to know what it was getting for its money at this important juncture.

Judge Holland made a decision that must have been personally painful. He rejected the criminal part of the plea bargain on grounds of insufficient restitution, stunning the negotiating teams. He criticized the proposed criminal fine and restitution in strong language. "I'm afraid these fines send the wrong message, which suggests spills are a cost of business which can be absorbed," Holland declared. With the criminal plea bargain rejected, the $900 million civil package sank soon after.

Insensitive comments by Exxon chairman Lawrence Rawl—who didn't even bother to attend the hearing—may well have been a factor in raising the judge's ire. A month before, in telling shareholders of the proposed settlement, Rawl had reassured them that they wouldn't feel it. He said the $900 million in civil penalties, to be paid over a ten-year period, would be deducted as tax write-offs, and that "the customer always pays everything." In other words, Exxon, which has annual revenues exceeding $100 billion, would feel no pain.

Whatever its cause, environmentalists across the nation praised Holland's decision. But Governor Hickel, on his way to Los Angeles to promote a water pipeline from Alaska, was at loss for words when reporters broke the news to him at the Seattle airport. Flustered, he insisted he would start renegotiations as soon as he returned. Five months later—only a week before Exxon's criminal trial was scheduled to begin in Anchorage federal court—Hickel had what was billed as a new deal. It was almost a duplicate of the failed plea bargain agreement, except that it provided $25 million more in criminal fines and restitution. Many questioned whether the new fines were enough to convince Judge Holland, and environmentalists called the modest ante an affront to the judge's strong mandate that penalties assessed against Exxon must send a message to rich corporate polluters.

On October 8, 1991, a platoon of lawyers, again trailed by a large corps of the media, once more filed into Judge Holland's courtroom. But this time the mood was entirely different.

No updated scientific report of damage assessment findings had been filed in advance of this proceeding, although the judge must have been aware of a *Los Angeles Times* story of the day before, reprinted by both Anchorage dailies, reporting secret projections from the government's economic studies that put the environmental damage from the spill as high as $15 billion. Paul Gertler had been yanked off the damage assessment management team just before the new deal was announced. He showed up in the courtroom as a spectator, unwilling to talk about his removal and saying only that he hoped Judge Holland would do what was best for the sound.

Exxon chairman Rawl also attended this time. Taking the defendant's box, wearing a repentant air and surrounded by a cadre of lawyers, the usually combative chief executive was a model of contrition. "Yes, Your Honor," he responded meekly to Judge Holland's routine questions on whether Rawl understood his rights.

Finally, Holland came to the single criminal charge brought against Exxon, a misdemeanor count for killing migratory waterfowl:

Holland: "How does Exxon plead?"
Rawl: "Guilty, Your Honor."
Holland: "Exxon is judged guilty."

The ritual was repeated with Augustus Elmer, fifty-one, president of Exxon Shipping Company, the subsidiary that operated the ill-fated tanker. He pleaded guilty for his firm to three misdemeanor charges involving killing waterfowl and violating the Clean Water Act and the Refuse Act. In return for the pleas, prosecutors dropped all four felony charges and the two remaining misdemeanor counts.

The formal pleadings out of the way, Holland then gave the lawyers a turn to address the adequacy of the total package. For the next hour, praises of Exxon and its generosity rang through the courtroom from unlikely sources. Barry M. Hartman, acting assistant U.S. attorney general for the Justice Department's Land and Natural Resources Division, pointed out that the proposed criminal payment was "off the charts," more than twice the amount collected by the government for all environmental crimes during the last seven years. He unfurled a corroborating chart to make his point. Charles DeMonaco of the environmental crimes section of the federal Justice Department said, "There is no question that Exxon has paid dearly for this oil spill." He stressed that the bulk of its future payments would help finish the job of restoring Prince William Sound. Charles Cole, attorney general for Alaska, said the state could put to better use the many millions it was spending to prosecute Exxon. By consenting to settle now, Exxon would save the state an estimated $25 million a year in legal expenses for many years down the road.

A smiling Judge Holland jotted notes. Then he had a few questions for Cole:

Holland: "I read in the newspapers, and you alluded to mutterings in the papers . . ."
Cole: "They make my blood boil."
Holland: "Simmer down . . . are you satisfied that the restitution money will get to where it is supposed to go, to the restitution of Prince William Sound? . . . [Will] it survive a challenge by the legislature to use it as it wants?"
Cole: "They won't have that luxury . . . as long as I am attorney

general, I guarantee that the money will be used for the restoration of Prince William Sound."

Next, Exxon's star lawyer, James Neal, a flamboyant Tennessean who had been a Watergate prosecutor, stepped to center stage. Shunning the traditional conservative attire of corporate barristers, Neal wore a rumpled tropical outfit. "I'm no silk-stocking lawyer," he began. "In fact, I'm an old rat who has been around the barn." Having properly humbled himself, Neal went on to detail the virtues of his client. He said Exxon already had paid out $3.5 billion for the spill cleanup and for settling some 12,600 related claims. Citing its $1 billion annual expenditures on environmental research and $40 million more for tanker traffic safety, he called Exxon a "great corporate citizen."

Judge Holland had one question. "Can the fines be passed on to the consumer?" With a wave of his hand, Neal deferred to Patrick Lynch, an Exxon lawyer from Los Angeles who was sitting beside Rawl. "We are a competitive industry," Lynch told the court. "We have no power to raise prices unilaterally."

Then Chairman Rawl reached for a microphone. "There is no question, I am sure we regretted this spill very much," he said. "We've done all we could do to get the spill cleaned up and deal with the people who came forward with claims." Judge Holland listened attentively and nodded. Then he declared a recess and said he would have a decision when he returned. As the standing-room-only crowd dispersed into groups in the courtroom and hallway, conversations could be heard about the dramatic change in the judge's demeanor.

When he returned to the bench, Holland announced he had received a fax message during the recess. It was from Representative Frank J. Guarini (D-N.J.), urging that the judge defer his decision until the House Ways and Means Committee could analyze the proposed settlement. "It's too late," Holland declared. "I accept and approve both the criminal plea agreement and the civil settlement."

Noting that he had reached his decision "only this morning," Judge Holland said he had had little time to prepare comments. Nevertheless, he went on to give a lengthy address. Holland said that two things had made the package acceptable. The criminal penalty was raised from $100 million to $125 million, "a twenty-five percent increase." And this time $100 million of the penalty was allocated toward restoration of Prince William Sound. Only $50 million had been so designated in the previous package, with the remaining $50 million to go to the federal government. Holland said he was "comfortable with the new information" provided at the proceeding. "What is now very clear to me," he said, "is that Exxon has been a good corporate citizen. It is sensitive to its environment obligations."

Acceptance of the plea bargain stirred intense reaction in the press. A skeptical *Anchorage Daily News*, the state's largest paper, noted Judge Holland's inconsistency and editorialized that the criminal fines packed too little punch. "There's not much hope of deferring corporate crime as long as people in charge can get off by signing a company check." Sue Libenson, executive director of the Alaska Center for the Environment, asked angrily, "What happened to the judge?" Governor Hickel proclaimed, "We are whole again."

Does $125 million up front and $900 million over ten years adequately compensate for the damage that the record oil spill did to Alaska? A definitive answer probably will not emerge until the government releases all the closely guarded findings from what may be the most massive scientific examination of a damaged environment ever undertaken. Soon after the oil spill, a small army of government biologists and contracted experts split into teams to work on fifty-eight scientific field studies of the damage to Prince William Sound. The studies were intended to produce the documentation that would enable the federal and state governments to present a claim for damages, as well as information on what might be needed to restore the injured resources. Nondisclosure clauses were written into the contracts to guard against leaks of secret data to the public.

Six agencies whose jurisdictions were most affected by the spill formed a trusteeship to finance the studies and to allocate any restoration funds recovered by legal action. The U.S. government had a trustee each from the Department of the Interior, which had control over federal parks, refuges, and fish and wildlife; the Department of Agriculture, whose Chugach National Forest borders the sound; and the Department of Commerce, whose National Oceanic and Atmospheric Administration monitors ocean mammals. The state had a trustee each from the Department of Fish and Game, the Department of Environmental Conservation, and the Department of Law. Each agency funded its own expenses in the scientific studies with assurance that they would be the first to be reimbursed by settlement money. "We set out looking for injuries," said Paul Gertler, the forty-year-old oil damage specialist who headed the management team. "We had to be prepared to document the injuries. In fact, the Department of Law went far and wide to recruit fifty peer reviewers. And we had to be prepared to work in a harsh environment. This was rugged territory. No roads. All work had to be done by boat, helicopter, and aircraft. Further, prespill data on the populations of various species had been skimpy or nonexistent."

After two field seasons of studies, a picture of the damage began to emerge. The most dramatic finding was the devastating effect upon the marine birds. Seabirds are particularly vulnerable in tanker spills. Oiled plumage causes loss of insulation and buoyancy, and birds die from hypothermia or drowning. Those that survive the initial impact often die later from ingesting the oil they frantically try to remove from their feathers. For days after the spill, the sound was littered with dead murres, sea ducks, bald eagles, loons, cormorants, and grebes, among other species. About 36,000 bird carcasses were recovered, but this was only a small percentage of the actual kill, the preliminary findings noted, because many stricken seabirds "floated out to sea, sank, were scavenged, were trapped and hidden in mass of oil and were not visible, or were buried under sand and gravel by wave action." Analyses provided by computer models estimated that the number actually killed by the spill was between 375,000 and 435,000 birds, although it could have been as high as 645,000.

The mortality of the common and thick-billed murres—the black seabirds with webbed feet and white breasts that resemble small penguins when standing on shore—alone was estimated at nearly 300,000. Hardest hit were the nesting colonies on Chiswell Island and Barren Islands, popular stops for nature tours. Oil spread into these major seabird nesting areas at the same time adult murres were congregating on the water near the islands in anticipation of the nesting season, killing 60 to 70 percent of breeding birds. The effect will be long lasting since murres breed slowly, generally laying one egg per year and only after they are four years old.

The decline of the harlequin duck, a colorful bird sought for sport and for its dramatic plumage, was so severe after the spill that the state had to impose a hunting ban in 1991 to spare the remaining breeding stock from annihilation. "Harlequins hatch along the streams and then bring their young into intertidal zones," Gertler explains. "They died by eating contaminated food—mussels, small shellfish, shrimp. No successful reproduction was found in the Prince William Sound area in 1991, two years after the spill." The ducks were reproducing normally in other areas.

Sea otters were still in trouble three years after the spill. While only 1,011 sea otter carcasses were recovered in 1989, preliminary studies estimate that between 3,500 and 5,500 animals—or nearly half of the sound's population—were killed directly by the spill. Of the 193 sea otters released from rehabilitation centers, 45 were fitted with radio transmitters, and only 14 of those were still alive. Adult otters and their young were dying at a higher rate in oil areas than in nonaffected areas. The otters continue to suffer from contaminated food as well as from a shortage of food, which is also attributed to the spill.

An estimated 200 harbor seals were killed by the spill. Their numbers declined 35 percent at oil sites, compared to 13 percent elsewhere. (Federal and state marine mammal experts have ruled out disease in the baffling general decline and now are studying overfishing as a possible cause.) "Severe debilitating lesions were found in the thalamus of the brain of a heavily oiled seal . . . and similar but milder lesions were found in five other seals collected three or more months after the spill," a study reported.

While Exxon press releases made much of the healthy return of pink salmon to Prince William Sound, the federal-state preliminary report noted disturbing signs about the long-range impact of the spill. The pinks that returned in great numbers were the hatchery fish whose environment was saved by the heroic acts of Cordova fishermen. The jury is still out at least until 1993, Gertler maintains, on the wild pink salmon, although in 1992 the Cordova fishermen reported that returning stocks of both hatchery and wild pink salmon were less than 50 percent of the expected yield.

It is still premature to link this poor season to the oil spill, but a long-term impact seems inevitable. Wild salmon did not shift spawning habitat after the spill and many deposited eggs in the intertidal areas of oiled streams, causing an inordinate number of salmon eggs to die. The studies found that in 1991, two years after the spill, the egg mortality rate in oiled streams kept increasing, to 40 to 50 percent, as compared to a mortality rate of about 18 percent elsewhere. An increase in deformities made the picture even grimmer. "Larvae from heavily oiled streams showed gross morphological abnormalities, including club fins and curved spines," the report states. In addition, gill parasitism and respiration rates were significantly higher in intertidal fish from oil sites.

Mussels and clams exposed to oil continued to show contamination three years after the spill. They metabolize hydrocarbons very slowly and therefore accumulate high concentrations. The study found that contaminated clams and other invertebrates "are a potential continuing source of petroleum hydrocarbons for harlequin ducks, river otters, sea otters and other species that forage in the shallow subtidal zone." The shellfish are also a major part of Native subsistence diets.

The least-known injury to the sound—and perhaps the most lasting—is the destruction of many varieties of sea-bottom life such as kelp, which is a basic food in the ecosystem. Fucus, the dominant intertidal plant, was severely affected not only by the oil but by the subsequent cleanup activities. While nature's cleansing processes continue to remove oil from the beaches, the displaced oil washes down into the biologically rich intertidal and subtidal zones and continues to harm marine and plant life.

A surprising extension of the damage was found farther inland. The spill adversely affected at least thirty-five archaeological sites, including aboriginal burial grounds and ancient dwellings. Oil covered many artifacts and killed the ryegrass cover on several of the sites, but the greatest danger to them probably was their discovery and looting by cleanup crews. Timely arrests and increased vigilance by Exxon stopped the abuse, but the disclosure of their location keeps the sites at risk. (To its credit, Exxon ordered shoreline evaluations to avoid damaging historic sites before cleanup crews could treat the beaches. Charles Holmes, a state archaeologist who helped pinpoint cultural legacies, said some artifacts went back four thousand years.)

Oil still hinders the pursuit of subsistence hunting, fishing, and gathering by the twenty-two hundred people who live in the fifteen Native villages within the spill-affected areas. The banks of the salmon stream fished by residents of Sleepy Bay on LaTouche Island are still blackened with oil. A clam-digger's shovel turns up a mixture of sand and oil. The situation is particularly tragic at Chenega Island farther up the sound, where an Alutiiq community of seventy people is fighting an uncertain battle to maintain its way of life. "The villagers used to take several hundred seals a year for their meat and oil," James A. Fall, a subsistence expert with Alaska Fish and Game, says. "After the spill, the seals started to disappear. There had been indications of a declining seal population beforehand, but Natives say the decline wasn't happening that fast."

The state monitors and tests the fish, shellfish, and animals in the sound and finds much that is now fit to eat. Fall tries to spread the word by newsletter and by speaking at village meetings. But Natives seem to place more credence in nature's signs than in the state's reassurances. Fall received this report from state Fish and Game subsistence division employees whose house-to-house interviews in Chenega also showed that subsistence harvests were down 57 percent from prespill averages:

> Indications from wildlife around them make the people very uncomfortable, and they are afraid to harvest subsistence food. An abnormal seal liver, ordinarily firm, was soft and runny. The arm of a starfish fell apart when pulled from the rocks. They have reported several dead eagles and sea gulls, a dead bear and a blind sea lion found during the past month, highly unusual occurrences prior to the spill.

All fifteen Native villages have joined in a class action suit against Exxon. While the physical damage to their environment is a major issue, the deterioration of their life-style is their most important claim for damages. "Without it, they lose their identity," anthropologists say. Once more, the spill has confronted the American legal system with an issue without precedent. No court before has set a value on the ruin of an Alutiiq life-style—or, for that matter, on any life-style.

The Alutiiq complaint is one of nearly three hundred lawsuits that remained unresolved after Exxon's plea bargain. Between twenty thousand and thirty thousand people have claimed damages against Alyeska, Exxon, and the other major oil companies in Alaska. The attorneys, of course, have been feasting. No less a personage than Melvin M. Belli, Sr., defender of the likes of Errol Flynn, Zsa Zsa Gabor, and Jack Ruby, advertised in the *Anchorage Daily News* one month after the spill that he had teamed up with a local law firm to pursue class action suits for damages. Four phone numbers were listed for interested callers.

Hundreds of workers who lost their jobs when Cordova canneries were shut down by the spill flocked to Belli and his local affiliate to file their claims. Other lawyers have started suits on behalf of Cordova fishermen and spotting pilots who maintain that the spill has robbed them of their livelihood. Landowners have joined the parade as well, on the grounds that their properties have been devalued. Timber cutters seek redress for alleged damages to their operations. A Native corporation links its Chapter 11 bankruptcy to the oil spill. Environmental groups are pursuing class action proceedings for funds to restore the sound, on top of what the state-federal trustees have already won.

Lloyd Miller, an Anchorage attorney, has been designated by the court to consolidate the cases to relieve congestion on the dockets. He expects trials to start by mid-1993, but indications are that Exxon and the other defendants will argue for extensions and use every other legal means to frustrate their accusers.

Many of the plaintiffs have visions of hefty awards. But the record suggests that sizable judgments, if any, may be a long time in materializing. The tanker *Amoco Cadiz* spilled 68 million gallons of oil off the coast of France on March 18, 1978. Plaintiffs won a judgment of about $125 million in damages, but not a dollar has been paid fifteen years later. The parties are still battling one another on cross-appeals, while only the lawyers grow richer.

Meanwhile, lines have formed for a piece of the billion-dollar settlement that Exxon is paying to the state and federal government. The court has said that the money must go to the restoration of Prince William Sound, but that is a vaguely defined purpose. Fifty million dollars has been put at the immediate disposal of the governor and legislature, while the remainder is to be spent on projects approved by the six appointed oil spill trustees. The opportunities for creative expenditures are many. Fears exist that Alaska's strongly prodevelopment governor may dip into the funds to build a road to the sound community of Cordova, which neither its citizens nor the legislature wants. The universities talk of a world-class research center. Two Native corporations want the state to use the money to buy their richly timbered land, or they will clear-cut it.

Paul Gertler and other experts who have been studying the sound feel it would be a mistake to apply restoration funds in any way except in protecting the habitat. "Removing foxes planted by man from islands with large bird nesting colonies would help bring back bird populations," Gertler says. "Buying parcels of timber to protect salmon streams and scenic shores is also compatible. But roads, docks, and even a research center would create only further environmental stress."

Nearly half of Prince William Sound is under federal jurisdiction. Therefore, all of America is entitled to be heard on how the billion dollars is to be spent over the next ten years. Unfortunately, the public cannot count on conscientious public servants like Paul Gertler to protect its interests.

Stanley Senner, a respected researcher who worked for more than two years as state restoration program manager for Alaska Fish and Game, has little confidence that the trustees will do what is best for the sound. He resigned in 1992 to take a job in the private sector, a decision undoubtedly hastened by his disillusionment with the trustees' priorities. Four of the six appeared committed to the maximum development of resources. One trustee had been a staunch defender of timber-cutting interests. Another had gained notoriety for advocating a proposal to cut twenty new roads and corridors into Alaskan wilderness, a nightmarish intrusion environmentalists dubbed "the spaghetti plan."

"Each trustee came in carrying his own agenda," Senner said. "Each carried a lot of baggage—the wish list of the person who named him to his job plus his own conception of resource management. As a group, they didn't know where they were heading or how to get there." Compounding the problem, each trustee has veto power over any restoration decision. But if by chance a trustee votes contrary to the wishes of the federal or state officer who appointed him, he can be replaced at will, as Paul Gertler discovered.

The bottom line is political decisions determined how much of the secret scientific data on the spill were released and when. And political decisions will determine how restoration money is spent.

CHAPTER 19
The New Battle

WALTER J. HICKEL, a lifelong Republican and one of the wealthiest men in Alaska, had tried five times to regain the governorship he had been elected to back in 1966. In the following twenty years, he ran in three primaries and waged write-in campaigns in two general elections, to no avail.

In 1990, at the age of seventy-one, Hickel said to hell with traditional election processes. He sat out the primary. Over the summer he grumbled privately with conservative friends about the wimpy platform of state senator Arliss Sturgulewski, the Republican nominee. Then, six weeks before the election, he made a bold and unexpected bid to retake the governor's mansion. In a maneuver plotted in Hickel's office, state senator Jack Coghill, a crafty sixty-four-year-old conservative from the interior district of Nenana who was seeking the lieutenant governor's office as Sturgulewski's running mate, agreed to defect from the Republican ticket. Only hours before the filing deadline in September, Hickel and Coghill had secretly constructed a new gubernatorial ticket with themselves on top under the banner of the obscure, ultraconservative Independence party. The party, of course, had already selected its candidates for governor and lieutenant governor in the primary, but the nominees, who knew they had no chance of winning, gladly removed their names. They were relieved to have such wealth and power at the head of their feeble, fund-starved party, whose greatest distinction up to then had been its espousal of secession from the United States.

Hickel dug deep into his pockets and set out to buy the election. The new ticket monopolized airtime and launched a blitzkrieg of newspaper ads. Voters were reminded of Hickel's humble beginnings on a Kansas farm and of his rise from poverty—from boxer to bartender, to builder, to governor, to secretary of the interior. And they heard again how he was fired by Nixon after he bravely urged the president to listen to the people's discontent over the Vietnam War. The campaign stressed that Hickel embodied the spirit of pioneer Alaska. "Remember the devastating earthquake of 1964?" the ads asked. While many were bailing out, Hickel stood amid the rubble and pledged to build a first-class, high-rise hotel on the very spot. (His five-star Captain Cook Hotel stands today as the centerpiece of the Hickel Investment Company, which controls $82 million worth of property in Alaska.)

Alaska's two Republican U.S. senators, Ted Stevens and Frank Murkowski, tried to convince Hickel to abort his campaign. He was splitting apart the Republican organization in Alaska, they argued, and practically handing the Democrats the governorship. Hickel did not waver. President Bush was out of the country, but Alaska Republican leaders persuaded John Sununu, the president's chief of staff, to apply pressure from the White House. In his call, Sununu not only appealed for party unity, Hickel says, but also hinted at unpleasant regulatory consequences for Hickel's latest major investment, a proposed natural gas pipeline from the North Slope. Hickel claims the veiled threat clinched his decision.

Hickel ran and won. It was the biggest election coup in Alaska's history. The new governor was elected by 39 percent of the voters, one of the lowest winning percentages ever. But Hickel succeeded in splitting the votes of his two major-party foes, who battled one another for the liberals and moderates. Democratic nominee Tony Knowles, the Yale-educated ex-mayor of Anchorage, and his running mate, Native leader Willie Hensley, received only 32 percent of the vote, despite Hensley's success in producing solid margins from Native communities. Moderate Republican Arliss Sturgulewski polled only 26 percent. Hickel concedes he spent $600,000 of his own money in the stretch. (His total campaign chest amounted to $800,000.) How he engineered the upset is an election oddity. Why he won, however, tells much about the people of today's Alaska.

However firmly he is etched in the tradition and legend of this rugged state, the independent man is becoming scarce in Alaska. The great majority of Alaskans today are spoiled and pampered citizens accustomed to receiving more money from the government than they pay to support it. As if annual disbursements from the Permanent Fund and an absence of state taxes were not enough, they have come to depend on state-subsidized housing mortgages and annual support from the public purse in many other places. One in five employees in Alaska is on a state or local public payroll. While the oil industry, through contracting out and buying early retirements, has been able to scale down the high wages and benefits paid to attract workers in the days when the pipeline was built, government employees' unions have locked the high pay and benefit standards established fifteen years ago into their contracts. Some Anchorage policemen clear more than $100,000 a year. Alaskan teachers, averaging $43,406 a year, are the highest paid in the nation. Public broom pushers pull down more than $30,000 annually. These salaries do not include pensions, vacations, health coverage, and free recreational facilities, benefits believed to be unmatched elsewhere in public employment. From the first day on the job, an Anchorage city worker is guaranteed a minimum of six weeks' paid leave each year, and state workers put in only a thirty-seven-and-a-half-hour week.

The notion that super salaries are needed to live in Alaska is rapidly becoming a myth. Dr. Scott Goldsmith, a professor of economics at the Institute of Social and Economic Research of the University of Alaska Anchorage, places the cost of living in Alaska's largest city at only 15 percent above the national average. True, winters are cold and long, but natural gas rates are among the lowest in the nation. Housing costs are comparable to those of a Pennsylvania suburb, and local real estate taxes are much lower. The cost of fresh food and medical care accounts for the greatest differential, but the cash payments from government plus the lack of an income tax or state sales tax nullifies much of that. Goldsmith's findings are supported by a 1991 federal study that determined that living in Alaska is now no more expensive than living in Washington, D.C., and recommended eliminating the 25 percent, untaxed extra-cost-of-living bonuses federal employees in Alaska receive.

Life has been so comfortable because royalties and taxes from oil pay 85 percent of the state budget. The money flows in from a resource few Alaskans see, and they have been treating it as though it will be coming in forever. Since 1981, the state has gone through about $35 billion of oil revenues, much of it dissipated as previous chapters have detailed.

But now the easy oil is already out of the ground at Prudhoe Bay. More than half of the field's estimated 10 billion recoverable barrels was tapped by the end of 1991. The remainder will be harder and costlier to recover. The long-feared decline of oil revenues has already started. Production peaked in 1988, and by the end of the 1990s oil output is likely to be half of the 1.9 billion barrels removed daily in 1991. Barring an extraordinary rise in oil prices, revenues will be cut at least in half. Alaskans have sufficient warnings of poorer times ahead, but they refuse to ease up on the good life. That's what got Walter Hickel elected. He convinced them he is a man with big projects and a track record of development who will find a way to keep Alaska rich after Prudhoe Bay runs out.

At age seventy-three, Hickel looks the part of a messiah. With a high forehead and fringes of gray hair, he talks to the masses with arms outstretched and imparts a feeling of earnestness. Away from the crowds, he skips rope to stay fit—a holdover from his boxing days, as is, perhaps, his tendency to become pugnacious when crossed. Hickel's main drawback, friends say, is that he often does not remember promises made from one week to the next. He frustrates reporters who try to pin him down on rationales for certain decisions by saying he listens to "the little man" within him. A second voice that seems to influence him is that of his wife, Ermalee, a tiny, gracious woman who is in the Barbara Bush mode except for her strong antiabortion views.

Soon after he was sworn into office, Hickel announced that Alaska would be operated henceforth as an "owner state." While he had trouble defining the concept, his cabinet appointments gave

Alaskans an early idea of what to expect. Hickel recruited an inner circle of boomers, many of them elderly colleagues in Alaska's prodevelopment think tank, Commonwealth North, which Hickel helped establish. They served notice that the state was in business to make a profit from its resources, which many interpreted as "Bull-dozers, start your engines."

To oversee the state's Department of Natural Resources, Hick-el picked oilman Harold Heinze, the former president of ARCO-Alaska and a longtime opponent of environmentalists as well as any new taxes on oil. In his new role, Heinze would oversee oil leasing, spill preparedness, and pipeline maintenance. These are the very areas at issue in the most controversial environmental battle facing the nation in this decade—the opening of the Arctic National Wildlife Refuge (ANWR) to oil drilling.

An arctic preserve about the size of South Carolina, ANWR sits on top of northeastern Alaska, abutting Canada. Its first lands were set aside as part of the nation's natural heritage by President Eisen-hower in 1960. Twenty years later, Congress nearly doubled ANWR to its present size with passage of the Alaska National Interest Lands Conservation Act, declaring it the most important wilderness sanc-tuary for arctic wildlife in the world. Many see this arctic expanse of tundra, rivers, and rolling hills as the last complex ecosystem in the world unaltered by man. It is also home to polar bears and more than one hundred thousand caribou of the ancient Porcupine herd that migrate there. The Native Gwich'in villagers, who live on an Athabaskan reserve to the south, depend principally on the caribou for food.

Heinze, however, is on record as calling ANWR "a flat, crummy place," and opening ANWR to oil drilling has the highest priority on the administration's development agenda. Hickel believes ex-ploration could produce a reservoir rivaling Prudhoe Bay, and oil-men say he may be right. A group led by Chevron USA drilled an exploratory well on Native corporation land adjoining ANWR in 1986 at a cost of $40 million. The findings have been kept secret, but the secrecy has only spurred the appetites of the developers. The discovery of a major new oil field would mean hundreds of new jobs in Alaska, and it would keep substantial oil revenues flowing into state coffers after Prudhoe runs out. Proponents also argue that it would alleviate the nation's dependency on foreign oil, though critics of drilling contend that its total projected reserves would fuel the nation for only two hundred days at the rate Americans consume oil.

The years of safe operations in Prudhoe provide the oil industry with a good case that it can explore and drill in the Arctic without ill effects on the caribou, polar bears, and environment, but the larger question remains unresolved. Should this rare and spectacular wilderness area be exploited before the nation has exhausted the alternatives?

For the moment, Hickel is restrained by Congress from developing ANWR. But he concedes no such limits in marketing another of the state's untapped resources, the estimated 35 trillion feet of natural gas reserves trapped beneath the North Slope. Great volumes of this natural gas are produced as oil is pumped from the wells at Prudhoe Bay. Because there is no way to market it, the gas is injected back into the ground. This is not a total waste. Reinjected gas keeps up well pressure, enabling oil to be pumped more easily and prolonging the life of marginal wells.

Hickel yearned early to find a more lucrative way to dispose of this great resource. Back in 1982, he founded the Yukon Pacific Corporation for the specific purpose of building a pipeline to transport the gas. The pipeline would be constructed along the 800-mile route blazed by the oil pipeline from Prudhoe Bay to Valdez, where a plant would chill the gas into liquid and tankers would haul it to market. But Hickel's grand scheme foundered. Natural gas is still inexpensive in the lower forty-eight and the market is soft. Sufficient customers could not be found to justify private investment in a pipeline system that could cost as much as $15 billion. And investors had no assurances that the oil companies would even sell the natural gas, since losing it would mean reduced pressure that poses risks to oil fields. Unwilling or unable to finance the project any longer, in 1988 Hickel sold control of Yukon Pacific to a rail conglomerate known as CSX. The *Anchorage Daily News* reported that the deal provided for Hickel to receive $10 million worth of stock in exchange for a willingness to "make himself available" to serve the project.

Hickel made bringing a big, new natural gas industry to Alaska a major plank in his campaign for governor. His supporters predicted the project would generate some $350 million in state revenue, create as many as ten thousand jobs, and help offset the nation's negative balance of trade. Few people had a chance to ask what such a pipeline would do for Hickel's personal fortune during his brief, frenetic campaign. The conflict-of-interest issue did not really surface until the fall of 1991, when the governor took off on a week-long, state-financed trip to promote the trans-Alaska natural gas pipeline in Japan and South Korea. (The trip occurred at the same time the ANWR debate was reaching a vote in Congress, leading even his supporters to wonder why Hickel was in Tokyo rather than lobbying contacts in Washington, as he had promised in his campaign.)

The governor came home to a lot of questions. The sales trip had focused interest for the first time on a conflict-of-interest complaint filed against Hickel in the spring by Juneau environmental activist Chip Thoma. Deeming the conflict issue too hot to handle on its own, the governor's own personnel board hired a special prosecutor to determine whether Hickel had violated the state's ethics act, which was passed in 1987. It was the first time an Alaska governor had had to face such a charge.

Hickel seemed to show no concern about being investigated by an agency in his own administration. "I have nothing to hide . . . voters knew about my involvement with this project when they elected me," Hickel told the press and then scolded them for dwelling on the negative. He reminded them that he had placed his Yukon Pacific holdings, estimated at 12 percent of the company's total stock, in a blind trust. But while in theory the trust manager was free to sell the stock without Hickel's knowledge, in practice the stock was virtually worthless unless a pipeline deal could be struck. And if a deal occurred, would any reasonable trustee deprive his client of millions in stock appreciation? Hickel said he would give all his Yukon Pacific stock to charity if that was what it took to get the pipeline built. The special prosecutor found no grounds to prosecute the governor, and there were only murmurs of dissent in Alaska.

The curious morality of the Hickel administration has been surfacing elsewhere as well. Shortly after the administration took office in 1991, Commissioner of Revenue Lee Fisher had sidestepped the state's bidding process and awarded a $125,000 contract to his old accounting firm. The job was to review tax assessments in a $3.8

billion dispute with the oil industry, and no written report was required. When the *Anchorage Daily News* broke the story of this extraordinary deal, Fisher replied testily, "If there was a procurement violation, then slap my wrists." The snide retort was too much even for Hickel. He fired Fisher as well as Commissioner of Administration Millett Keller, who approved the contract, but with high praise for their overall performance.

Nevertheless, Hickel has proceeded just as ruthlessly himself when established procedures have got in his way. Obsessed with building roads across Alaska, he authorized the state's Department of Transportation to start bulldozing a route through the Copper River valley to the isolated fishing village of Cordova at the southeastern edge of Prince William Sound. The legislature had turned down funds for a road to Cordova the spring before, and no permits had been granted to begin the work. The citizens of Cordova themselves had voted five times against such a road. But the administration raided maintenance funds to start the moving earth and simply ignored the permit process—more or less a replay of the building of the Hickel Highway.

Outraged, an assistant attorney general named Ellen Toll fired off a letter to the press without identifying herself as a lawyer working for the state. She protested "the governor's misappropriation of highway maintenance funds" and described what she saw at the scene: "Workers were bulldozing trees, dirt, and rock into the Copper River and its tributaries, destroying a historic railroad right of way and filling in wetlands adjacent to the river, all without any kind of surveying, planning, or permits from any other agency, state or federal." Toll was fired right after her piece appeared. But this time Hickel had created a greater problem for himself.

Earlier in the year, a couple of diehards in the Independence party, still resenting the way the elected nominees of their party were shoved off the ballot, launched a recall campaign against Hickel and Coghill, claiming the pair had been illegally placed on the party ticket. The recall was going nowhere, but the Copper River road controversy suddenly gave it new ammunition. Citing the reckless bulldozing along a fine Alaska salmon river, the powerful Alaska chapter of the Sierra Club voted to join in the recall. "Our policy disagreements with the Hickel administration are not grounds for recall," a spokesperson wrote. "What is ground for recall is the governor's repeated disregard for the rule of law and for his responsibility to execute the legislature's decision in the manner intended."

None of the eight daily newspapers in Alaska backed the recall. In fact, the press has been generously supportive of most of Hickel's big projects, even though reporters privately talk about several of them as being "off the wall." There has been little editorial derision of Hickel's scheme to sell water to drought-stricken California via a 1,700-mile pipeline along the Pacific Ocean floor, his $100 million plan to buy Fire Island in the Cook Inlet and create a deepwater port to serve Anchorage (despite the lack of evidence that sufficient customers would pay to use it), or his commerce commissioner's attempt to fund a feasibility study of a plan to beam electricity to Eskimo villages via satellite from distant generating plants. This last scheme is dubious both technologically and economically, but so far editorial ridicule has not been forthcoming.

At the time Hickel took office in 1990, a fierce newspaper war was raging in Anchorage, home of the state's two statewide papers. An unlikely new general in the battle was Bill Allen, fifty-two, the former welder who headed Veco, and the most militant supporter of the oil industry in Alaska.

Veco had been the prime contractor in cleaning up Prince William Sound, taking an estimated $800 million of the approximately $2.5 billion Exxon spent. Late in 1989, Allen plowed some of his cleanup profits into buying the failing *Anchorage Times*, the venerable prodevelopment voice of Alaska. He acquired the paper for $6.2 million from the aging Robert Atwood, just as it appeared that the liberal, proenvironment *Anchorage Daily News*, owned by McClatchy Newspapers Inc. of California, would sink it.

This was only the latest strange twist in a war that had been raging for decades. Up until the last ten years, the *Times*, an afternoon paper edited and published by Atwood since 1935, had dominated the state. A seemingly feeble challenge by the *Daily News* started early in the 1970s. At that time, Larry and Kay Fanning, a Chicago editor and his new wife, who was recently divorced from a scion of the Marshall Field family, sought a new life in Alaska. They bought the *Daily News*, a morning paper that was published in a rundown shop and had a third the circulation of the *Times*. The *News* struggled, Larry Fanning died, and his widow went through a traumatic joint operating venture with Atwood that ended in suits and countersuits. At one point, Kay Fanning had to appeal for public donations to save the *News*. Finally, in 1979, she made a desperate trip to California. Stressing the big oil boom about to start in Alaska, she convinced the McClatchy newspaper chain to buy her paper.

As the oil boom brought prosperity and new readers into Alaska, McClatchy poured millions into upgrading the *News*. The company moved it into a new, $10 million publishing showplace with modern color presses, and would later invest $30 million more. Howard Weaver, an Alaskan who'd been educated at Johns Hopkins, gave the paper editorial conscience. The *News* fought unbridled development, corruption, and abuses of human rights. By 1983, it had surpassed the *Times* in circulation. By 1990, it had won two Pulitzer Prizes and a national reputation for excellence. At home, however, it began to hear charges of arrogance, contempt for development, and undue sniping at the oil industry.

"It's a war," Bill Allen declared soon after settling into the *Times* publisher's office in 1990. He proclaimed he would keep alive a strong prodevelopment editorial voice, enhance "fair and accurate" news coverage, and run his "biased competitor" out of town. Allen began pouring millions into energizing his newspaper. He doubled the news staff, brought in proven professionals to run it, and installed new computers and modern color processing capabilities. And in his most daring move, he converted the *Times* to a morning newspaper and began competing head-on with the *News*.

Aided by the oil companies, which bought his paper by the thousands to distribute to offices and schools, the *Times* bounced back from a low of twenty-nine thousand readers to more than forty-six thousand. It was still short of the *News*, which had sixty-one thousand subscribers, but it was the first newspaper in the nation that switched to morning publication in the midst of a competitive battle and gained readers.

The battle was costing both papers a profit, but dividends surfaced in unexpected places, notably for Governor Hickel in Juneau. The new vigor of the *Times* seemed to move the *Daily News* to soften its often harsh opposition tone. Instead of sledgehammer editorials, criticism of the governor and his operations was increasingly relegated to a humor column written by Mike Doogan.

On June 2, 1992, even as his editors were still hiring new staff and expanding news coverage, Allen ran out of steam and money. He called his staff together that afternoon and announced that the battle was over. He had just sold the paper to McClatchy, which would shut it down immediately.

A novice at publishing, Allen had clearly underestimated the difficulty of changing readers' habits. It was an expensive lesson. He lost about $37 million in the venture, and although both parties refused to divulge the sale price, sources in the profession say McClatchy picked off Allen's paper with all its new computers and other additions for only $8.5 million. McClatchy shut the *Times* down the day after it bought the paper, putting about 471 people out of work.

In his Veco headquarters, which feature a wild West motif, Allen rationalizes the loss as he stretches his muscular frame and places his cowboy boots on the arm of a cream-colored leather sofa. "I wouldn't have done it if I had any idea it would cost that much," he says, "but this hasn't been a failure. After the oil spill, I believe the *Daily News* would have annihilated the oil industry, run it out of Alaska. We forced the *Daily News* to become a fairer paper. And, you know, I did negotiate a deal that guarantees that our view will be heard in the *News* for a long time."

As part of the complicated deal, the *News* agreed to let Allen buy half the op-ed page of each issue for ten years. He retained three editors who fill the space with prodevelopment editorials, many of which are rebuttals to or criticisms of the *Daily News* editorial positions or news coverage. "It's like a choke chain on a dog," Allen says. "If it gets too nasty, we pull the collar." Allen concedes it is costing him $500,000 a year to maintain the arrangement, which exists at no other newspaper in the country.

Some say one casualty of the newspaper war is a permanent softening of the *News*'s position. To the great disappointment of its longtime fans in the environmental movement, it, along with every other newspaper in the state, has come out with a flat endorsement of oil drilling in ANWR. The papers reflect the prevailing mood among Alaskans, but the press is unlikely to have much influence in the controversy in any case. The decision to drill is up to Congress, because ANWR is federal land. And Congress has several larger issues to consider in the battle besides the interests of Alaskans.

President Bush made the opening of ANWR a key part of his proposed national energy policy. Congress, which had protected the refuge from development since 1980, did relent to a point: In 1991, it showed a willingness to consider removing the ban on drilling as part of a comprehensive plan to meet the nation's pressing energy needs. The debate comes down to an overarching question about national priorities: Is the national interest best served by opening up the last potential major oil field in the country now, or by first attacking the wasteful ways that Americans use oil?

President Bush, who set a poor example by roaring around in a gas-guzzling Cigarette boat during his summer trips to Maine, showed little patience for giving greater consideration to conservation or other alternatives for meeting energy needs. He wanted ANWR opened to oil drilling—period. Proposed tougher conservation measures did not make much of an impact on the automobile industry, either. Detroit has opposed any energy program that would require it to produce autos and light trucks averaging 40 miles per gallon by the year 2000. When the auto industry refused to budge in a 1991 showdown, Congress refused to release ANWR.

But this is simply a lull in the battle. It would be a national tragedy if the decision were made on ANWR's development potential only. The country needs a rigorous national energy policy—with incentives for conservation and a commitment to specific plans to develop alternative sources—before it tears into the nation's last few pockets of wilderness.

And what is the future for Alaska?

Despite its touches of sophistication, Alaska in many ways seems to be a state out of control. Pickup trucks sport gun racks in their rear windows. Drugstores stock rifles and revolvers a few feet away from the aspirin. And it is a rare edition of a daily paper that does not have at least one story involving a shooting. A teenage driver is shot and killed by another for cutting in front on him on the Glenn Highway. A sport fisherman is shot to death and his companion wounded for disturbing the wilderness experience of a trio who resent them tying their boat to the same U.S. Forest Service mooring. An apartment manager in Anchorage drills three bullets into a tenant because he refuses to turn down his radio.

But if Alaska seems to attract the wanton and the reckless, it is this same spirit of freedom that also lures the enterprising adventurers and the committed caretakers. The state has a fascination about it. People come up for a year and stay for a lifetime. The latest émigrés look as diverse as the landscape. But take away the utter freeloaders and those running away from trouble, and the newcomers are surprisingly the same. Each in his own way feels invigorated by the awesome power of Alaska—its formidable mountains, its rugged shoreline, and its wildlife. Man is unprotected here; this is nature's world.

The oil industry underestimated the widespread respect for this rare quality. The *Exxon Valdez* oil spill incurred international outrage, and the oil industry's long pattern of deceptions in a remote part of the world caught up with it. The money it spent spying on critics and financing surveys to rebut environmental complaints would have been spent more effectively on simply fulfilling its promises and correcting its deficiencies.

But has the oil industry learned any lessons? Improved spill contingency plans have blossomed, true, but spill prevention has not improved. Too many old and poorly maintained tankers remain on the seas to satisfy this nation's insatiable thirst for oil. An appalling 425,000 tons of oil were spilled by tankers by the time 1991 was half over, according to the Tanker Advisory Center in New York. That is more than twice the amount spilled during all of 1989, the year of the record spill in Prince William Sound. On December 3, 1992, the Greek tanker *Aegean Sea* crashed onto the rocks outside the fog-shrouded harbor entrance at La Coruña, Spain, spilling 21.5 million gallons of crude oil, again almost twice the amount lost by the *Exxon Valdez.* And in 1993, before the year was a week old, the Liberian tanker *Braer,* on its way to Canada with a full load of oil from Norway, lost control during stormy seas and hit the rocks off the Shetland Islands. It lost its entire cargo of 24.6 million gallons, topping even the *Aegean Sea* spill. The Oil Pollution Act of 1990, inspired by the Prince William Sound disaster, mandates that any tanker built after June of that year must have a double hull if it is to operate in

U.S. waters. However, under terms of the same act, phaseouts of existing single-hull tankers will not start until 1995 and will take place over a period of the next twenty-five years.

Alaska is often called the nation's last storehouse of minerals. It has large untapped deposits of coal, lead, zinc, and natural gas, and some old-timers insist that the largest gold lodes remain to be discovered. Besides keeping stricter watch on the oil industry, the state will have the growing burden of seeing that the way these resources are extracted is compatible with maintaining Alaska's character.

Alaska already faces problems from the lack of development conscience among the relatively new Native corporations. Some 44 million acres of Alaska are in their custody. More than a few Native corporations are harvesting their prime timberlands as though there is no tomorrow. They clear-cut scenic hillsides because no government regulation says they cannot. But then they turn to government to rescue them when the money runs out. Special-interest legislation has already saved several from bankruptcy, and they have returned to legislation to insulate them from hostile takeovers. Without such protection, the weaker corporations will surely lose their land. But they ask, in essence, what no ordinary business would dare to demand—protection from their own imprudent management. As crucial as it is to help Natives keep their land, Congress must find a way to do so without imposing a double standard on American corporations.

Since the first gold rush, a "cut it, dig it, and cart it away" philosophy has governed the way Alaska treats its natural resources. That is no longer tolerable. The beauty of the land is itself a natural resource, and protection of it must also be part of Alaska's development. "We need to redefine the meaning of the last frontier," says Lee Gorsuch, a longtime Alaskan who is dean of the School of Public Affairs at the University of Alaska Anchorage. "We must enhance its mystical value and emphasize compatible industries—tourism, research, state-of-the-art oil technology. We must develop it frugally while conserving its value."

Fundamental to this conflict is the question: What does Alaska mean to the nation and the world? Aside from the ANWR issues, environmentalists feel that as undeveloped land grows scarcer, it is more important than ever to preserve Alaska. Meanwhile, Alaskans reject the idea that their fate is to be park managers.

It is important for all Americans to be aware of what is happening to this nation's truly Great Land. The rule of law, not political expediency, should determine how it develops. We have seen how easily the law has been manipulated at times in Alaska. As hungry special interests attempt to redefine the last frontier for their own advantage, the temptation of such tactics is great.

Alaska is embroiled in a war over its future, in which a multitude of factions competes. How this war is resolved will reflect how America views this incomparable part of the world.

Acknowledgments

Any acknowledgment must start with thanks to the state of Alaska. I came in 1987 for a year, to teach and fish. The Alaska mystique hooked me, so I stayed to write this book. I am now in my sixth year of residence and can't bear to leave.

Ideas for the book started to develop early in my first year as Atwood Professor of Journalism at the University of Alaska Anchorage, two years before the oil spill. Some chapters were inspired during an interdisciplinary journalism and economics class, which I team-taught with Dr. Steve Jackstadt, a talented economics professor. For class discussions, we interviewed public officials who discussed their spending excesses with no remorse, developers who recounted unconscionable exploitation, and newspaper people who had no qualms about discussing any fraud on their beat.

The university is the home of the Institute of Social and Economic Research, directed by Lee Gorsuch. I quickly discovered the vast expertise of those occupying the cells at this think tank. I debated history with Tom Morehouse, statistics with Scott Goldsmith, and theories with Lee Huskey. I talked politics with Gorsuch and received expert guidance in their fields from Professors Matt Berman,

Terry Smith, and Steve Colt. Elsewhere in the university, the perspectives of anthropologist Steve Langdon and librarian Ron Lautaret were immensely valuable, as were the insights of historian Steve Haycox and the research of Professor Arsenio Rey.

The oil companies were cautious but usually helpful. Ronnie Chappell, communications specialist at ARCO, arranged interviews in the Prudhoe Bay oil fields, and Charlie Elder, Jr., British Petroleum's sage of the pipeline era, steered me to excellent sources. Gene Rutledge, an oil researcher and writer, made available his mountain of data, and Jack Roderick, ex-borough mayor and author, shared his wisdom and even continued the dialogue on the tennis courts.

Francine Taylor put me in touch with Fairbanks friends who told me about life on the North Slope during pipeline days. Tim Bradner, an Alaska business analyst without peer, and his wife, Shelah, who worked in Barrow, introduced me to their circle of contacts. Almost to a person, public officials obliged with their versions of government deeds and misdeeds, many of which they themselves have published.

Countless people briefed me and welcomed me into their homes as I traveled the length and breadth of the state. I cannot mention them all, but I would be remiss not to acknowledge the assistance of Robin Harrison and attorney Tom Lohman in Barrow. Dr. Kurt Vitt and Walter Larson in Bethel, Riki Ott and Mark Madrid in Cordova, Kenneth Holbrook in Yakutat, Richard Fineberg in Juneau, Nancy Freeman in Kodiak, Jack Sims in Prince of Wales, and Chris Westwood in Ketchikan.

Newspapers are the first version of history. Anchorage was lucky to have had two good dailies. I read them, clipped them, and spent many hours poring over issues from the days before microfilm. While I developed friends on both papers, none were more helpful than the library staffs, headed by Mary Dye at *The Anchorage Times* and Sharon Palmisano at the *Anchorage Daily News*. As she has during my entire writing career, my wife, Nancy Jordan, also a journalist, served as critic and helpmate.

I am grateful for the philanthropy of Robert Atwood, who endows the Atwood chair, which brought me to Alaska. My thanks also to the University of Alaska, which provided me with an office at the consortium library and an appointment as writer in residence after my two-year turn as Atwood Professor had expired. Sylvia Broady, the chair of the journalism department, looked after my needs even as my teaching duties diminished. Barbara Sokolov, the library director, gave me a key to the library, enabling me to work at odd hours.

So many at the university helped, but in true academic tradition, there are no strings attached. This book is solely mine, and I alone am responsible for its content.

Friends back east, knowing my desire to fish untraveled waters, ask how I manage to survive in a state whose dangers in the wild are renowned. It is a joy, I tell them. But that calls for a final word of appreciation. My thanks to the bears I met at the Akwe River and at Thorne Bay. It was kind of them to regard me as a wayward oddity rather than a morsel from the mainland.

Index

Adams, Al, 140, 145
Adams, Jake, 126, 127–28
Adams, Marie, 124
Aegean Sea, 267
AFN (Alaska Federation of Natives), 69–75, 94
Agnew, Spiro T., 84
Agricultural Revolving Loan Fund, 103, 105, 114
Agricultural Stabilization and Conservation Service, U.S., 108
agriculture, 99–109
  ARRC investments in, 113
  barley production, 102–3, 107–8
  dairy farms, 100, 102, 103–7, 108–9
  state spending on, 102, 108, 118
AHFC (Alaska Housing Finance Corporation), 119–21
Ahmaogak, George, 131, 132
Alascom, 162, 163, 169
Alaska:
  agricultural programs of, 99–109
  annual budget for, 10, 100
  banking in, 121, 154–55, 156, 159, 160–61, 182, 183
  capital investment programs in, 111–22, 179–89, 210
  climate of, 13, 15, 49, 50–51, 110, 123, 205
  congressional delegations from, 70, 213
  cost of living in, 123–24, 257–58
  dividends paid to residents of, 119, 122, 257
  environmental groups in, 36, 60, 130, 247, 252, 262–63; *see also* environmental issues
  government corruption in, 126–50
  gubernatorial elections in, 102, 105, 113, 133, 207, 215, 227, 255–57, 258
  labor unions in, 88–93, 162–74, 212
  lawless image of, 93, 266–67
  local taxing authority in, 207, 208
  military bases in, 32, 87, 155
  mining in, 15, 23–24, 32, 77, 87, 113, 186, 268
  in national security strategy, 31–32, 48
  Native land rights in, 21–22, 41, 61–75, 175–76, 185, 189, 197
  newspaper war in, 263–65
  1986 recession in, 160–61, 172–73
  no-tax policies in, 118, 119, 122, 194, 257–58
  oil revenues paid to, 10, 46, 54, 98, 100, 101, 110, 118, 119, 153, 194, 209, 258, 259
  partisan politics in, 101, 127, 129, 133, 136, 208, 210, 212–15, 255–57
  permanent savings fund created by, 10, 118–19, 122, 257
  population of, 10, 14–15, 119, 153, 158–59, 191
  racial segregation in, 66–67, 176
  railroads in, 78, 79, 87, 103, 155, 205
  rain forests of, 186–89
  roads in, 59–60, 68–69, 78, 178, 196, 205, 253, 262
  Russian interests in, 66, 77, 124, 175
  statehood of, 24, 31–32, 43, 47, 51, 64, 65, 68, 207, 208
  state legislature of, *see* state legislature, Alaska
  tourism in, 113, 180, 191, 194–201
Alaska (Michener), 131
Alaska Advocate, 46
Alaska Board of Game, 204
Alaska Center for the Performing Arts, 153
Alaska Conservation Group, 60

Index

Alaska Development Company, 25
Alaska Federation of Natives (AFN), 69–75, 94
Alaska Fish and Game, 42–43, 247, 251, 253
Alaska Hospital, 166, 168, 170, 171
Alaska Housing Finance Corporation (AHFC), 119–21
Alaska National Bank of the North, 182, 183
Alaska National Interest Lands Conservation Act (ANILCA) (1980), 202, 203, 259
Alaska Native Claims Settlement Act (1971), 75, 175, 189, 201–2
Alaska Oil Spill Commission, 236–37
Alaska Pacific University, 157
Alaska Power Authority, 117
Alaska Renewable Resources Corporation (ARRC), 111–14
Alaska Scouting Service, 28
Alaska State Bank, 121
Alaska State Council on the Arts, 153–54
Alaska-Yukon Magazine, 26
alcoholism, 177
Aleutian Islands, 31
Aleut Native corporation, 189
Aleuts, 15, 76, 124, 175
Allen, Bill, 213–14, 231, 263, 264–65
Alutiiq, 251
Alyeska Pipeline Service Company, 89, 90, 92, 93, 94, 97
  Exxon Valdez spill and, 228, 229, 231, 237–40, 252
Amalgamated Development Company, 26
Amerada Hess, 59
American Newspaper Publishers Association, 32
American Petroleum Institute, 43
Amoco, 64
Amoco Cadiz, 252
Anchorage, development in, 32, 119, 151, 152–61, 166, 179
Anchorage Cold Storage, 169, 171, 172, 173
Anchorage Daily News, 46, 92, 131–32, 133, 168–69, 202, 247, 252, 260, 262–65
Anchorage Independent, 34
Anchorage Times, 16, 23, 28, 29, 30, 31, 38, 46, 71, 84, 98, 171, 263, 264–65
Anderson, Robert O., 56, 80, 81
Angapak, Nelson N., 178
ANILCA, *see* Alaska National Interest Lands Conservation Act
ARCO, 154, 213, 214, 259
  see also Atlantic Richfield
Arctic National Wildlife Refuge (ANWR), 51, 240, 259–60, 261, 265–66
Arctic Slope Native Association, 68, 75
Arnett, Ray, 38
Arnold, Bill, 81
ARRC (Alaska Renewable Resources Corporation), 111–14
Arsenault, Lennie, 133, 136
arts funding, 153–54
Associated Press, 218
Athabaskans, 15, 175, 259
Atlantic Refining Company, 55, 56–57
Atlantic Richfield, 57–58, 59, 80, 81, 152, 153
  see also ARCO
Atuk, Richard, 183
Atwood, Evangeline, 31, 47
Atwood, Robert B., 28, 33
  Anchorage development and, 152
  newspaper published by, 263, 264
  oil industry and, 16, 29–30, 38, 40, 42, 44–46
  statehood promoted by, 31–32, 47

bald eagles, 235
barley production, 102–3, 107–8
Barnes, Ramona, 118
Barrow:
  capital improvement projects in, 126, 128–33
  corruption prosecutions and, 132–33, 134, 136, 137–50
  mayoral elections in, 125, 127–28, 131
  Native community of, 124–26
  oil wealth in, 125–26, 152
  remoteness of, 123–24
Barrow Utilidor, 126, 130–31, 144–45, 146
Bartlett, E. L., 70
Baskin, Harvey, 99–100, 103–4, 106–7, 108–9

Index

bears, 123, 197, 202, 259
Bechtel, Ed, 233
Bechtel Corporation, 89
Beck, Dave, 164, 167
Bell, George, 182
Belli, Melvin M., Sr., 252
Bering, Vitus, 77
Bering Straits Native Corporation, 181–84, 189
Berlin, Rudy, 57
Bethel, 178
Beynon, L. R., 237
BIA, *see* Bureau of Indian Affairs
Bishop, Sam, 227–28
Bishop, William C., 34, 35–38, 47
bison, 108
H. W. Blackstock Company, 129, 132, 133, 145, 146, 147, 148–49
Blomfield, Susan, 86–89, 95–96
bonds, industrial, 80–81
Bonner, Herbert, 42
bowhead whales, 124
Boyd, Randy, 121–22
Boyko, Edgar Paul, 40
Bradshaw, Thornton F., 57
Braer, 267
Breeden, Loretta, 195
Bristol Bay native corporation, 189
British Petroleum (BP), 211
  Iranian facilities of, 28, 50, 51
  North Slope explorations by, 50–51, 54, 57, 58–59
  oil spill preparedness of, 237–38
  pipeline project and, 80
  salaries paid by, 154
  Sohio subsidiary of, 152
broadcasting industry, 185
Brock, Joseph P., 139–40
Brower, David, 83, 84
Brower, Eugene:
  fiscal mismanagement by, 126–27, 129–32, 137–38, 140
  in mayoral elections, 128–29, 131
  North Slope prosecutions and, 139, 140–41, 143–44, 146, 148–50
Brown, Boyd, 48, 49, 53, 59
Brown, Hal, 136
Bullock, Edith, 21
Bureau of Indian Affairs (BIA):
  business loan guaranteed by, 180
  oil lease deals made by, 63–64
  as paternalistic bureaucracy, 62–65
  schools run by, 18–19, 20, 68
Bureau of Land Management, 16, 39–41, 45, 67–68
Burns, James, 147, 149
Bush, George, 256, 266
Business Week, 56

Calista Native Corporation, 178–81, 184, 189
Canada, oil discovery in, 27
caribou, 52, 175, 191, 202, 203, 204, 259
Carr, Helen, 167, 172
Carr, Jesse L., 91–92, 93, 162–73
  death of, 172, 212
  dictatorial style of, 162–64, 168–70, 171
  government investigations of, 93, 164, 165
  national leadership and, 167–68, 171–72
  negotiating tactics of, 91–92, 164, 169
  union finances and, 91, 164, 165, 166–67, 168, 171, 173
Carr, J. Randy, 88, 91
Cesarini, Raymond, 228
CFAB (Commercial Fishing and Agriculture Bank), 114–16
Chan, Ronald G., 144
Chevron, 28–29, 209, 259
Chiappone, Rick, 151–52, 161
Chugach Eskimos, 76–77, 189
CIRI (Cook Inlet Region Inc.), 68, 185–86, 189
Cities Service, 55, 56
Clark, Ramsey, 74
Clifford, Clark, 45
Clinton, Bill, 137
Coast Guard, U.S., 217, 218, 220, 222, 223–24, 225, 228, 229, 238
Coffman, David, 147
Coffman and White, 139, 144
Coghill, Jack, 255, 262
Cole, Charles, 183, 245–46
Commercial Fishing and Agriculture Bank (CFAB), 114–16
Commonwealth North, 259
Congress, U.S.:
  Alaskan delegation in, 70, 213

Index

Congress, U.S. *(cont.)*
on Alaskan statehood, 32, 47
on ANWR drilling, 260, 265–66
environmental issues in, 36, 41–42, 82, 240
farm subsidies enacted by, 108
Native land claims and, 71, 73, 75, 175, 184, 185, 189, 268
NOL tax break permitted by, 184, 188
pension funds regulated by, 165–66
pipeline project approved by, 84–85
wildlife refuges created by, 202, 259
Constantine, Alec, 61
Constantine, George, 61
Cook, James, 77
Cook Inlet, oil development in, 61, 64
see also Kenai Peninsula, oil industry on
Cook Inlet Region Inc. (CIRI), 68, 185–86, 189
Cooper, Stephen, 147, 148
Copeland, Tom, 231–33
copper mining, 24
Cordova District Fishermen United, 231
COREXIT 9527, 229
Cornett, Don, 230–31
Cousins, Gregory, 220, 221–22, 224
Cowper, Steve, 213, 215–16, 227–28, 242
crab fishing, 115
Crane, Ed, 116
Croft, Chancy, 208–9, 211, 212, 225, 238
Crosby, Bing, 23
cultural development, 153–54
Curran, Harold, 130, 134

dairy industry, 100, 102, 103–7, 108–9
Dankworth, Ed, 117, 134, 210, 214–15
D'Arcy, Christine, 153–54
Dash, Sam, 136
Davis, Mark, 147
Dawson, Kent, 206
Decker, Thomas, 148, 149
Defense Department, U.S., 31
Delozier, Mark L., 223–24
DeMay, Peter, 81, 92
DeMonaco, Charles, 245
Dennis, LeRoy, 53
Department of Environmental Conservation (DEC), 228, 235, 237, 239, 240, 247
Desert Horizons Country Club, 167, 168, 170, 173
Dischner, Lew:
background of, 127, 165
Barrow construction programs and, 128, 129, 130, 132, 143
political campaigns and, 127–28, 129, 131, 133, 145
prosecution of, 138, 140, 141–46, 148, 149, 150
dog teams, 207
Doogan, Mike, 265
Douglas, William O., 45
Dupre, Bob, 128

Economic Limitation Factor (ELF), 215–16, 227
Edwardsen, Charles, 68, 69, 75
Egan, William, 51, 80–81, 127, 207
Egan Civic and Convention Center, 153
Ehrlichman, John, 71
Eisenhower, Dwight, 15, 32, 34, 42, 44, 55
Elder, Charles, Jr., 95
Elmer, Augustus, 245
Employee Retirement Income Security Act (ERISA) (1974), 165–66, 167, 170
Environmental Defense Fund, 82
environmental issues, 267–68
Alaskan groups on, 36, 60, 130, 247, 252, 262–63
of *Exxon Valdez* spill, 218, 234, 235, 236, 241–42, 243, 244, 247–51
for hunting and fishing, 194, 196–204
of oil industry controls, 36–37, 40–44, 79, 208–9, 216, 236–40
on pipeline project, 78–79, 82, 83, 84
for timber industry, 186–89, 268
of wildlife protection vs. oil interests, 30, 36, 41–44, 47, 259–60
Environmental Protection Agency (EPA), 82, 84
Erickson, Gregg, 210
ERISA, *see* Employee Retirement Income Security Act
Eskimos:
cultural values of, 19
gift-giving ritual of, 144
two major language groups of, 14–15

Index

see also Inupiats; Natives, Native peoples; Yupiks
Espe, Arnold, 158–59
Evans, Neil, 147, 148
Exxon, 29, 78
Exxon Shipping Company, 225, 230, 241, 245
Exxon Valdez oil spill, 217–54
cleanup efforts for, 228–36, 240
costs of, 231, 233, 234, 235, 236, 240, 241–44, 246–47, 263, 267
environmental damage from, 218, 234, 235, 236, 241–42, 243, 244, 247–51
extent of, 218, 222, 228, 230, 240, 241
first reports of, 217–18, 228–29
fishing industry and, 230, 231–34, 235, 242, 252
government restoration settlements for, 241–47, 252–54
investigations of, 219–26, 236–37
jobs created by, 231, 234
Natives affected by, 230, 235, 236, 250–52
private damage claims and, 246, 251–52

Fahd, King of Saudi Arabia, 156
Faiks, Jan, 214–15, 216
Fairbanks:
Anchorage development vs., 152
oil strike celebrated in, 58
pipeline employment in, 86–87
Fairbanks News-Miner, 133, 163, 227
Falkenstein, Thomas, 224
Fall, James A., 251
Fanning, Kay, 168, 263–64
Fanning, Larry, 72, 263–64
Farm Credit Administration, 114
Farrell, Martin, 140, 150
Federal Bureau of Investigation (FBI), 93, 137–40, 141, 146, 165
Federal Deposit Insurance Corporation (FDIC), 161, 183
Fidalgo, Salvador, 77
Finegold, Laurence, 143
Fischer, Victor, 102
Fish and Wildlife Service, U.S., 31, 36, 235
Fisher, Lee, 261–62
fishing, fishing industry, 190–204
environmental damage and, 194, 196–201, 249–50, 252
Exxon oil spill and, 230, 231–34, 235, 242, 252
legal restrictions on, 190–91, 192–94, 201, 202, 203
Native interests and, 179, 181, 193–94, 201–2
recreational, 191, 194–95, 196, 197–200
salmon, 32, 62, 116, 175, 179, 190–95, 196, 198–201, 249–50
state investment capital for, 111, 112–13, 114, 115–16
trapping methods used in, 17, 18, 192, 193, 198, 200, 201
Fitzgerald, James M., 146
Fitzgerald, Joseph, 80
Fluor Corporation, 89–90
Fortas, Abe, 45
Fosdick, Dean, 218
Fowler, Geoffrey, 144–45, 147
Frampton, George, 135
France, Dan, 197–98, 201
France, Mary, 197
Frank, Richard, 69
Friends of the Earth, 78–79, 82, 83
From the Rio Grande to the Arctic (Jones), 47
fur trade, 13–14, 15, 77, 113, 124, 195

Gamache, Peter, 141, 142, 148
geese, 178
Gertler, Paul, 243, 244, 248, 249, 253, 254
gill nets, 192, 193, 201
Gilson, George, 79
Gilson, Phil, 200–201
Gittins, Thomas, 140, 145
Glacier Bay National Monument, 65
Goldberg, Arthur J., 74–75
Goldberg, Robert, 74
gold mining, 15, 23, 26, 77, 87, 113, 268
Goldsmith, Scott, 257–58
Gorsuch, Lee, 268
Gorsuch, Norman, 133, 135, 136
Goss, Connie, 234
Groh, Cliff, Sr., 70, 72
Gross, Avrum M., 93
Gruening, Ernest, 43, 66–67, 70
Gruenstein, Peter, 46

Index

Gryc, George, 51
Guarini, Frank J., 246
guide services, 198, 199, 200, 201, 202–3, 204
Guy, Mark, 218

Haida Indians, 65, 69, 116
Hamel, Charles, 239–40
Hammock, Dave, 217
Hammond, Jay, 206, 208
  agricultural program of, 102–3, 105, 108, 109, 111, 118
  capital investment programs of, 111, 112, 113, 117
  separate accounting repealed by, 210–11
Hanrahan, John D., 46
Hansen, Jan, 91
Harding, Warren, 50
Hardy, Oliver, 23
Harlan, John Marshall, 45
harlequin ducks, 248–49, 250
Harrison-Chilhowee Baptist Academy, 20
Hartman, Barry M., 245
Havenstrite, R. E., 25
Hawkins, Scott, 159
Hazelwood, Joseph, 219–21, 222, 223–26
Heinricks, Ray, 21
Heintz, William, 164, 169
Heinze, Harold, 259
Hensley, Jessie, 19
Hensley, John, 17, 18
Hensley, Priscilla, 13, 17, 18, 19, 20
Hensley, William, 18, 19
Hensley, Willie, 13–22
  education of, 18–22, 65, 67
  Native corporation headed by, 186
  native land rights championed by, 21–22, 64–69, 70, 73, 74
  political career of, 67, 68, 70, 256–57
  village life of, 13–14, 17–18, 19
Herbert, Charles, 51
Hickel, Ermalee, 258
Hickel, Walter J., 58, 161
  Exxon prosecution and, 243, 244, 247
  gubernatorial campaigns of, 102, 255–57, 258
  on Native land claims, 70, 71–73
  pipeline project and, 82
  prodevelopment policies of, 59–60, 78, 258–59, 260–63
  public criticisms of, 261–65
  as secretary of interior, 44, 70, 71, 72–73, 78, 82, 83
  subsistence rights and, 203
Hickey, Dan, 134–35, 136, 137, 165
highway system, 59–60, 68–69, 78, 178, 196, 205, 253, 262
Hill, Mason, 34
Hiner, Tim, 199–200
Hinson, Charles, 144
Historic Anchorage, Inc., 153
Hoffa, Jimmy, 167
Hohman, George, 134
Holland, H. Russell, 242, 243, 244–47
Holmes, Charles, 250
Home Oil of Canada, 59
Hopson, Eben, 125–27, 128, 130, 131
House of Representatives, U.S., 32, 42, 55, 75
  see also Congress, U.S.
housing, 119–22, 154, 257
Humble Oil:
  early Alaskan exploration by, 29
  North Slope operations of, 54, 55, 78
  pipeline project and, 80, 89
  see also Exxon
Hundorf, Roy M., 185
Hunt, H. L., 27–28
Hunt, Nelson Bunker, 28, 152
Hunter, Celia, 60
hunting, 13–14, 191–92, 201–4, 249
  see also fur trade
Hutchins, Floyd, 232
hydroelectric power projects, 117–18

Iarossi, Frank, 230
IBEW (International Brotherhood of Electrical Workers), 162
Igtanloc, Irving, 139, 140, 148–49, 150
Igtanloc, Myron, 140, 150
Independence Mine, 23
Independence party, 255–56, 262
Indian Claims Commission, 175
Iniskin Drilling Company, 25
Institute of Social and Economic Research, 102, 216, 257
Interior Department, U.S.:
  Exxon oil spill damage and, 247

Index

Hickel's tenure at, 44, 70, 71, 72–73, 78, 82, 83
Native land rights and, 64, 65
oil lease policies of, 15–16, 34, 36, 39–45, 64
pipeline project and, 82, 83–84
Internal Revenue Service (IRS), 138, 144, 188, 194
International Brotherhood of Electrical Workers (IBEW), 162
International Brotherhood of Teamsters, 162–74
Alaskan membership of, 91, 164–65, 168, 173
democratic efforts within, 168–70, 171, 172
finances of, 91, 164, 165–67, 168, 170–71, 172–73
Labor Department action against, 168, 170–71
national leadership of, 167–68, 171–72
organized crime and, 92–93, 165, 167, 168
pension fund for, 91, 165–66, 167, 168, 170, 171
political influence of, 127, 165
International Union of Operating Engineers, 88–89, 92
International Whaling Commission, 124
Inupiats, 15, 124–26
see also Eskimos; Natives, Native peoples
Iran, 28, 50, 51, 97, 100

Jackson, Henry, 73, 74
Jackson, J. R., 54
Jacobs, Locke, Jr., 16, 30, 33, 37, 39–40, 46
Jamison, Harry, 53, 54, 57, 58
Janice N., 232, 233
Jaworski, Leon, 136
Jessup, Don, 57
Johanssen, Neil, 196–97, 199
Johnson, Lyndon B., 80
Johnstone, Karl, 225, 226
Jones, Charles S., 34, 37, 43, 44, 47, 55, 56
Jones, Maureen L., 221, 222
Jones, Stan, 133, 134, 135
Jones, W. Alton, 55–56
Josephson, Joseph P., 215
Juneau, isolation of, 205
Justice Department, U.S.:
antitrust actions by, 55–56
corruption case prosecuted by, 138–50
Exxon prosecutions and, 242

Kagan, Robert, 221, 222
Keim, Lori, 96–97
Keller, Millett, 262
Kellogg Oil, 33
Kelsey, Bob, 78
Kelsey, John, 77–78, 79–81
Kelso, Dennis, 228, 239
Kenai, as oil boom town, 195–96
Kenai Peninsula:
environmental damage on, 196–201
Kenaitze land rights to, 41
oil industry on, 29, 30, 31, 32, 34, 37–38, 41–46, 195–96
wildlife refuge on, 30–31, 34, 35, 36, 41–44, 47
Kenai River, 194–201
Kenaitze Indians, 41, 195
Kennecott Copper Corporation, 24
Kennedy, John F., 55, 56, 62
Kennedy, Robert F., 55–56
Kent, P. E., 58
Kito, Sam, 206
Kleess, Guy, 221
Klondike gold strike, 15
Knowles, Tony, 157, 256
Koniag Native corporation, 189
Kotzebue region, 14, 67–68, 186
Kunkel, James R., 222–23

Labor Department, U.S., 168, 170–71, 173
labor unions, 88–89, 90, 91–93
see also International Brotherhood of Teamsters
Lacovara, Philip, 136
Larkin, Wayne, 148, 149
Larson, John, 130
Laurel, Stan, 23
Lawn, Dan, 223, 228, 237, 238, 239, 240
Leffingwell, E. de K., 50
Lehr, Ron, 117–18
Lekanof, Flore, 69
LeMay, John A., 170

Index

LeResche, Robert E., 111, 113–14
Libenson, Sue, 247
Life Systems, 182
Linton, Jack, 120–21
Littleton, Wayne, 112
Loeffler, Karen, 141–42, 144, 148, 149
Los Angeles Oil and Refining Company, 33
Los Angeles Times, 92, 93, 244
Lost Frontier (Hanrahan and Gruenstein), 46
Louisiana, Lacassine refuge in, 42
Loussac Library, 153
Lynch, Patrick, 246

McAniff, Roger, 137, 145–46, 149
McCarthy, Glenn, 27, 89
McClatchy Newspapers, 263, 264, 265
McCool McDonald Architects, 139
McCord, Emil, 61–62, 63, 64, 69
McCutcheon, Stanley, 62, 63–64, 65, 70
McDonald, Allen, 147
McDowell, Sam, 203
McGovern, George, 73
McKay, Douglas, 15, 34, 36, 41–42, 44
Malaspina Glacier, oil lease refused for, 40–41
Mangus, Marvin S., 52, 55, 57
Marathon, 34, 46
Marshall, Tom, 51
Matanuska Maid creamery, 104, 105, 107
Mathisen, Carl, 128–29, 130, 132, 138, 141–46, 148, 149, 150
Medred, Craig, 202
Meekins, Russell, Jr., 101, 121, 134, 210
Mello, Chris, 129, 130, 134, 149
Michener, James, 131
Middleton, R. Collin, 183–84
Miller, George, 239
Miller, John C. "Tennessee," 48, 49, 53, 57
Miller, Lloyd, 252
Miller, Richard A., 19, 20
Miller, Terry, 121
Mineral Leasing Act (1920), 82, 84
MMCW, 139–40, 144, 145, 147
mobile-home financing, 120–22
Mobil Oil, 40, 59
Moore, Bruce, 135
Moore, Martin B., 180–81
Moore, Roseleen "Snooks," 115
moose, 30–31, 42–43, 47, 202
Moquawkie Reservation, 62, 64
Morgan, J. P., 24
Morse, Wayne, 42
Morton, Rogers C. B., 83, 85
Mossadegh, Mohammed, 28
Mount Edgecumbe, 20, 68
Mt. McKinley Meats, 104
Mull, Gil, 52–53, 54, 57–58
Munn, James, 148
Murkowski, Frank, 182, 183–84, 213, 256
Murphy, Ed, 219–20
murres, 248
Museum of History and Art (Anchorage), 47, 153
muskrat trapping, 13–14, 15

NANA Regional Corporation, 186
Nasser, Gamal Abdel, 28
National Bank of Alaska, 91, 159, 166
National Environmental Policy Act (1969), 82, 84
National Marine Fisheries Service, 194, 201
National Moose Range, 30–31
  see also Kenai Peninsula, wildlife refuge on
National Oceanic and Atmospheric Administration, 236, 247
national petroleum reserves, 50
National Transportation Safety Board (NTSB), 219, 220, 221–25
National Wildlife Federation, 36, 41
Natives, Native peoples:
  activist organizations formed by, 67, 68–69
  business ventures of, 116, 179–89, 253, 268
  cash economy vs. traditional subsistence culture of, 10, 11, 14, 62, 95, 124–26, 177, 179, 191
  education of, 18–19, 20, 68, 126, 127
  electoral voice of, 69, 125, 257
  employment difficulties of, 94–95
  enslavement of, 124, 175
  environmental issues and, 187–89, 268

Index

Exxon oil spill and, 230, 235, 236, 250–52
fishing rights of, 193–94, 201–2
health problems of, 75, 177–78, 179
as hunters, 13–14, 94, 124, 201–4
land claims of, 21–22, 41, 61–75, 175–76, 185, 186, 189, 197
North Slope corruption case and, 138, 139, 140, 144, 149, 150
oil leases and, 21–22, 41, 52, 61–65
pipeline project and, 79, 82, 83, 94–95
poverty of, 62, 75, 95, 178
rapid acculturation effects on, 176–78
reservations for, 62, 175
segregation practices and, 66–67, 176
subsistence rights claimed by, 191, 202–4, 250–51
substance abuse among, 95, 177–78
sudden wealth acquired by, 64, 95, 124–26, 176
tax break for, 184–85, 186, 187–89
thirteen regional corporations established for, 175–76
three main groups of, 14–15
see also specific Native groups
natural gas, 25, 27, 46, 256, 260–61, 268
Navy, U.S., oil exploration by, 50
Neal, James, 246
net operating losses (NOLs), tax loophole for, 184–85, 186, 187–89
New York Times, 25–26, 59, 80, 83
Niemeyer, Mike, 179
Nixon, Richard M., 42, 75, 85
cabinet appointments by, 44, 71, 83, 256
oil industry support for, 55, 71
wage freeze instituted by, 162
Watergate affair and, 135, 136
NOLs (net operating losses), tax loophole for, 184–85, 186, 187–89
Nome, 181
Nordstrom, 154
North Slope:
geological mapping of, 50
Native community of, 68, 75, 124–26, 138, 149
size of, 49–50
wildlife of, 52
North Slope, borough of, 125, 126–27, 138
see also Barrow
North Slope, oil industry on, 48–60
access routes for, 49, 53, 59–60, 78–79; *see also* pipeline
environmental impact of, 53–54, 59–60
first big strikes of, 57–59
largest North American oil field at, 58–59
lease sales for, 51–52, 54–55
Native wealth gained from, 75, 124–26
weather conditions and, 49, 50–51
North Slope Constructors, 132, 143
Northwest Alaska Native Association, 67
Notti, Emil, 68–70, 71–74
NTSB (National Transportation Safety Board), 200, 221–25

Odom, Milt, 169
oil industry, Alaskan:
Anchorage headquarters for, 152–53, 156
climatic conditions and, 26, 49, 50–51
decline expected for, 258
early exploratory efforts in, 23, 24–35
environmental controls on, 36–37, 40–44, 79, 208–9, 216, 236–40
federal land rights obtained for, 34, 36–45
financial markets affected by, 58, 59
first big strike in, 37–47
geological conditions for, 26–27, 34, 50, 57
on Kenai Peninsula, 29, 30, 31, 32, 34, 37–38, 41–46, 195–96
Mideastern oil industry vs., 28, 34, 84
Native land rights and, 21–22, 41, 52, 61–65, 75, 175–76, 185
in northern tier, 48–60, 75, 78–79, 118, 124–26, 258
oil prices and, 26, 34, 97, 98, 100, 118, 121, 159, 211
political concerns of, 11, 101, 124, 206–16, 227
seismic studies for, 36–37, 52–53
state revenues generated by, 10, 46, 54, 98, 100, 101, 110, 118, 119, 153, 194, 209, 258, 259
tanker safety controls and, 78–79, 208–9, 216, 219, 236–39, 267–68

Index

oil industry, Alaskan *(cont.)*
tax policies and, 101, 125, 206–11, 215–16, 227
wildlife preservation vs., 30, 36, 41–44, 47, 259–60
see also oil leases; oil spills; pipeline
oil leases:
acreage restrictions on, 29
Anchorage investment group for, 16, 29–30, 31, 32–33, 37, 44–46
federal policy on, 15–16, 34, 36, 39–45, 64
private profits from, 45–46
rates per acre for, 15, 30, 54–55
state administration of, 51–52
top-filing claims on, 44–45
Oil Pollution Act (1990), 267–68
oil spills:
burning of, 229–30
dispersants for, 229
immunity from prosecution for, 225, 226
pipeline planning and, 83, 84
preparedness for, 209, 228, 236–40, 267
worldwide incidence of, 252, 267–68
see also Exxon Valdez oil spill
Olympic, 145
Organization of Petroleum Exporting Countries (OPEC), 84, 97
Ott, Riki, 230
otters, 77, 124, 234, 235, 249, 250
Owens, Tom, 116
Owens Drilling Company, 116
Owsichek, Ken, 203

Palmer, Bob, 102
pamaq, 144
Pan American Oil Company, 64
Pan American Western Petroleum, 33–34
Papetti, Ralph, 176
Parker, Randy, 88
Parker, Walter B., 80, 236–37
Parker, William, 119
Patton, Edward L., 89, 92
Paulson, Forest J., 115
Pearson, Drew, 72
Pelly, Thomas M., 40–41
Peratrovich, Robert, 69
permafrost, 50–51, 59
Persian Gulf War (1991), 156
Pessel, Gar, 52–53, 54, 58
Pettyjohn, Fritz, 133, 138
Phillips Petroleum, 27, 29, 34, 59
pipeline, 79–98
completion of, 97, 98
construction of, 81, 85, 89–90
cost of, 80, 85, 90, 97
daily flow of, 90, 98
environmental issues and, 82–84, 216
oil companies involved in, 59, 80, 81, 89
planning of, 79–85, 89
route of, 97–98
workers on, 86–96, 165
Place in the Sun, A (Wills), 238
polar bears, 123, 259
pollock fishing, 201
Pollock, Howard, 70
Pope, Douglas, 142–43, 144
Powell, Larry, 187
Presser, Jackie, 172
Prince William Sound, 78
oil spills and, 84, 217, 218; *see also Exxon Valdez* oil spill
wildlife of, 76, 77, 186–87, 248–51
Project 80s, 153
Prudhoe Bay, oil field at, 10, 50, 51–60, 75, 78, 118, 258
see also North Slope, oil industry on; oil industry, Alaskan
Pruhs, Dana, 143

Quaker Oats Company, 188

Rabinowitz, Jay, 67
Racketeer Influenced and Corrupt Organizations (RICO) Act (1970), 138, 145
Rader, John, 207–8, 212, 225
railroads, 78, 79, 87, 103, 155, 205
rain forests, 186–89
Rasmuson, Elmer S., 29, 30, 33
Rasmussen, Brent, 141
Rawl, Lawrence, 243–45, 246
Rayburn, Sam, 32
Reagan, Ronald, 148, 243
Red Dog zinc mine, 186
Reimer, Terrence, 179
Rice, Julian, 83–84

Index

Richardson, Claire, 217
Richardson Highway, 78, 103
Richart, John, 53
Richfield Oil:
  Atlantic merger with, 56–57
  in early Alaskan oil exploration, 27, 33, 34–38, 41, 43–44, 45, 46, 47
  financial history of, 33–34, 55
  North Slope operations of, 49, 52–55, 56–58
  see also Atlantic Richfield
RICO (Racketeer Influenced and Corrupt Organizations) Act (1970), 138, 145
Rock, Howard, 69
Roderick, John, 28, 29, 32
Rogers, Roy, 68
Rogstad, Kenneth, 129, 132, 133, 141, 147–48
Roosevelt, Franklin D., 31, 43, 65
Rose, Dave, 122
Rowan, Irene, 188–89
Rowland, Mark C., 183
Rudd, Lisa, 135
Russia, purchase treaty with, 21, 66
Russian trappers, Natives enslaved by, 124, 175
Ryan, Ben, 53
Ryan, Irene, 26–28, 46
Ryan, Pat, 27

Sabah, Jaber al-Ahmed al-, 156
Sackett, John, 121
Salamatof Seafoods, 112
salmon, 32, 62, 116, 175, 179, 190–95, 196, 198–201, 249–50
  see also fishing, fishing industry
Schenk, L. George, 191
Schwartz, Chuck, 47
seals, 17, 249, 251
Seaton, Fred A., 42, 43–44
Seiser, Virgil, 39–41
seismic exploration, 36–37, 53–54
Selman, Charles, 53, 57
Senate, U.S., 32, 36
  Native land claims studied by, 71, 73, 74–75
  see also Congress, U.S.
Senner, Stanley, 253
setnets, 192, 193, 200, 201
Settlers Bay, 181, 184
Shee Atika, 187–88
Sheffield, William, 105, 107, 113, 133–34, 135–37, 139, 159, 160
Shell Oil, 34
Shepphird, George, 34
Sheraton Hotels, 180, 184
Shively, John, 135, 136, 160–61
Sierra Club, 262–63
Silver Bullet, 223
Sinclair Oil, 55, 56, 57
Sinnett, Robert J., 172
Sitnasuak, 183, 184
Slama, Jack, 172, 173
Sohio, 152
Spaan, Mike, 138–39, 141, 142, 144–45, 146–48
Spear, William, 111
Speer, Edgar, 81–82
Spencer, David L., 36
Spencer, Hal, 148
Spokane Bank for Cooperatives, 114, 116
Standard Oil of California, 27, 29, 34, 45
Standifer, Bill, 61
Starr, Red, 233
state legislature, Alaska, 205–16
  corruption charges and, 134
  fishing regulated by, 192–93
  oil interests and, 11, 101, 206–16, 227
  permanent savings fund established by, 118–19
  revenue surge and, 100–101
  Sheffield impeachement considered by, 136–37
  on subsistence rights, 203
  on tax policy, 101, 118, 207–11, 215–16
  venture capital programs of, 111–12, 117, 120–21
steel industry, 81–82
Stevens, Francis M., 62–63
Stevens, Ted, 84, 138, 184, 213, 256
Sturgulewski, Arliss, 115, 116, 206, 212, 255, 257
subsistence rights, 191, 202–4
Sullivan, George, 157
George M. Sullivan Sports Arena, 153
Sun Oil, 105
Sununu, John, 256
Supreme Court, U.S., 45, 210

Index

Susitna dam project, 117, 118
Swanson River, Richfield drilling site near, 35, 37–38, 41
Sweet, John, 57
Swift, Ernest, 36

Talbert, G. Bruce, 137–38, 139, 141, 146, 149
Tallman, James, 45
Tanana Chiefs Conference, 69
tanker safety controls, 78–79, 208–9, 216, 219, 236–39, 267–68
Tanner, Jack, 148
TAPS, *see* Trans-Alaska Pipeline System
tax policy:
  on municipal jurisdiction, 207
  Natives favored by, 184–85, 186–89
  no–tax tradition in, 118, 119, 122, 194, 257–58
  oil industry and, 101, 125, 206–11, 215–16, 227
Taylor, James C., 50
Teamsters, *see* International Brotherhood of Teamsters
Tepa, 112
Thoma, Chip, 261
Thometz, Kreg, 204
Thornburgh, Dick, 241, 242
Tillion, Clem, 192, 194, 208, 212
timber industry, 32, 114, 116, 186–89, 252, 253, 268
Tlingit Indians, 65, 69
Toll, Ellen, 262
Tolman, Frank, 34
Tongass National Forest, 65, 186–88
Topel, Marc, 148, 149
top-filing, 44
Torrey Canyon, 220
Trans-Alaska Pipeline Authorization Act (1973), 84–85, 88
Trans-Alaska Pipeline System (TAPS), 80, 82, 89
  see also pipeline
Treaty of Cessation (1867), 66
Tri-Leasing, 143
Truman, Harry S., 197
Trustees for Alaska, 130
Tundra Times, 69
Tyee Lake, hydroelectric project at, 117–18
Tyoneks, BIA oil leasing ended by, 61–65, 69

Udall, Stewart L., 70, 71, 72
ulus, 17
Union Oil, 29, 34, 46, 59
United Bancorporation, 160–61
Unocal, 29
U.S. Geological Survey, 34, 50, 51, 52
U.S. Steel, 81–82

Valdez:
  grain terminal built in, 103, 107
  history of, 77–78, 218
  as pipeline terminus, 78, 79–80, 81, 152, 237–39
  population of, 76, 77
  see also Exxon Valdez oil spill
Valdez Dock Company, 77, 78
Veco International Inc., 213–14, 231, 233, 234, 265
Venezuela, oil leases in, 37
Vietnam War, 83, 256
von Braun, Wernher, 84
von der Heydt, James, 171

Wackenhut, 240
Walker, Ann, 177–78
Walker, Jack, 39, 40
Wall Street Journal, 160, 188
Warren, Earl, 45
Waterer, Tom, 112
waterfowl, 178, 235, 244–45, 248–49
Watergate hearings, 135, 136, 246
Watt, James, 38
Weaver, Howard, 46, 264
Wells, Arlo, 172
Wester, Wilbur, 30, 32–33
whaling, 17–18, 124, 126
"What Rights to Land Have the Alaska Natives?" (Hensley), 65
"Where Did All the Billions Go?," 216
White, Bill, 131–32
White, Emerson, 173–74
White, Peter, 139, 144, 147
White, Tom, 25
White Alice, 162, 163

Index

Wilderness Society, 78–79, 82
wildlife:
  hunting restrictions and, 191–92, 201–4
  of North Slope, 52
  oil interests vs., 30, 36, 41–44, 47, 259–60
  of Prince William Sound, 76, 77, 186–87, 248–51
  see also specific wildlife species
Wildlife Management Institute, 41–42
wildlife refuges:
  ANWR, 51, 240, 259–60, 261, 265–66
  drilling leases granted in, 30, 36, 41–44, 47
  on Kenai Peninsula, 30–31, 34, 35, 36, 41–44, 47
  subsistence rights in, 202–4
  in Yukon Delta, 178
William IV, King of England, 77
Williams, Roy, 167–68
Williams, Tom, 210, 211
Wills, Jonathan, 238
Wilson, Woodrow, 62
Windeler, Ron, 162–64, 166, 168, 169, 170, 171, 172
wolves, 202, 204
Wright, Robert S., 59
Wright, Sande, 105–6, 107
Wyatt, Bill, 79

Young, Don, 213
Yukon Delta Wildlife Refuge, 179
Yukon Pacific Corporation, 260, 261
Yupiks, 15, 178, 180–81
  see also Eskimos; Natives, Native peoples

Zamarello, Patricia, 160
Zamarello, Peter, 155–60, 161
Zappa, Anthony, Jr., 16